ANALOG INTEGRATED CIRCUITS

ANALOG INTEGRATED CIRCUITS

DEVICES, CIRCUITS, SYSTEMS, AND APPLICATIONS

J. A. CONNELLY

Contributing Editor
Associate Professor of
Electrical Engineering
Georgia Institute of Technology

Prepared from contributions by the
Engineering Staff of
Harris Semiconductor
A Division of Harris Corporation
Melbourne, Florida

A WILEY-INTERSCIENCE PUBLICATION

John Wiley & Sons

New York • London • Sydney • Toronto

Published by John Wiley & Sons, Inc.

Library of Congress Cataloging in Publication Data:

Connelly, Joseph Alvin, 1942–
  Analog integrated circuits.

  "A Wiley-Interscience publication."
  Bibliography: p.
  Includes index.
  1. Integrated circuits. I. Harris Semiconductor
(Firm) II. Title.

TK7874.C66   621.381'73   74-20947
ISBN 0-471-16854-8

Printed in the United States

10 9 8 7 6 5 4 3 2 1

To Mary Nelle
and to
Tim, Jeff, and Stephen

CONTRIBUTORS

James Beasom
Richard Jenkins
Robert Webb
Leo Enriques
Donald Jones
Ernest Thibodeaux

FOREWORD

One of the greatest potential risks in preparing a book on any twentieth-century technology is that by the time the words fall into place on paper they may no longer represent the real "state of the art." I fear too many books do a disservice to the reader, who digests the fare only to discover very quickly that he has consumed old meal garnished with buzz words to make it appetizing.

Thus it was with considerable reservation that I agreed to participate in this venture. My most serious doubts concerned *balance:* could a comprehensive work on any aspect of microelectronics maintain "equal time" integrity? Could the authors keep their vaunted technology in its proper place? Could they collectively avoid the temptation to ballyhoo their state of the art, taking undue and perhaps undeserved credit for only a passing figure in a kaleidoscope of change? It is neither improper nor altogether immodest to take pride in the genuine progress of the last decade, provided the accumulated failures and successes are catalogued intelligently and this knowledge forcefully articulated.

Even more important, I wanted to be sure that the authors could devote sufficient attention to the fundamentals and philosophy of integrated circuits. Could they give the reader the solid footing he needs to grapple with the complexities of our technology?

The answers to these questions formed the basis of participation: we would proceed only as long as the authors maintained their balance—only as long as they laid their foundations solidly before erecting the structure.

I am satisfied that *Analog Integrated Circuits* has met the criterion of balance faithfully. Although the book is confined to design, philosophy, manufacture, and application of analog devices, the reader will find here, perhaps for the first time in one work, the essentials of a highly amorphous craft. In a very real sense, we have chronicled an industrial happening as precisely as imperfect words will permit.

Understandably, I am proud of the combined efforts of Harris Semiconductor authors. These men have made a profound contribution to the library of industrial and technological knowledge. Integrated circuitry is still a new field; we learn new things every day, yet the fundamentals and philosophy that have guided this industry since its birth only a dozen years ago hold inexorably. At the very least we have carefully recorded these basics; at most, we will inspire younger and bolder minds.

Either way, our industry will be all the better.

D. R. SORCHYCH
Vice President
and General Manager
Harris Semiconductor

Melbourne, Florida
August 1974

PREFACE

Within the short span of approximately one decade, analog integrated circuits have reached a level of maturity at which basic fabrication technologies, circuit designs, and key applications have become reasonably well defined. It is now quite appropriate to draw together in one coherent source many of the designs, philosophies, and applications that have become "the basics" of analog ICs.

The development of the first "linear" integrated circuit operational amplifiers fathered a near-infinite collection of "linear" devices that fill numerous manufacturers' catalogs. The vast number of both general and special-purpose "linear" ICs currently available makes it difficult to accurately classify them as "linear." In this work, we have chosen to use the broader meaning associated with "analog" for better classifying the scope of topics covered.

Harris Semiconductor has made significant contributions to the development of analog ICs. The documentation of many of these previously unpublished contributions together with the associated background information and philosophies presented herein serve as a convenient vehicle to enable interested individuals to acquire a broad and timely working knowledge of analog integrated circuits.

This book was written to provide a unified and qualitative understanding of the principles utilized in the fabrication, circuit design, system operation, and application of analog integrated circuits. The work serves as a useful reference for engineering and science specialists such as fabrication engineers, circuit designers, system analysts, applications engineers, and others who need to broaden their acquaintance with analog integrated circuit topics. Educators, students, and technicians who already possess a basic understanding of solid-state devices and circuits also will find the work particularly helpful for its coverage of many useful and practical applications of analog ICs.

Many engineers who are recognized experts have contributed their particular insight into their chosen specialty areas. These contributions have been edited and arranged into a unified and complete presentation of analog ICs. Chapters 1 and 2, which were mainly prepared by James Beasom, describe the basic fabrication steps involved in the processing of all monolithic integrated circuits. Distinctions between processing philosophies for digital and analog elements are discussed, and the standard analog IC process is described. Specialized analog processes utilized to achieve unusual element characteristics such as improved isolation, high-frequency performance, and super-β devices are discussed and illustrated. State-of-the art technologies are described, together with their accompanying limitations and predictions for further technology refinements.

Chapter 2 presents the techniques and models most frequently used by the device engineer to characterize analog elements. Typical results obtained with these characterizations are compared with results obtained experimentally. The advantages and disadvantages of integrated resistors, capacitors, and transistors are described and compared with the characteristics of their discrete counterparts. Computer utilization for circuit analysis and design is described by Richard Jenkins.

Chapter 3 is divided into the two parts—basic bipolar circuit structures and specialized circuit structures. My major contribution was to the first part, with assistance from Richard Jenkins in describing basic gain and output stages. In the specialized structures, Leo Enriques provided the material for FET circuits, admittance cancellation techniques, and positive-feedback networks. Robert Webb described the modified differential amplifier, analog mixing circuits, resistor network configurations; with Ernest Thibodeaux, he prepared the material on analog switches. The philosophies incorporated into the circuit structures of Chapter 3 are described and related to the processing technologies and device characteristics presented in the previous two chapters. These circuits then serve as building blocks for the construction of analog devices treated in the following chapters.

Donald Jones and I treat the functional operation and circuit organization of voltage comparators and operational amplifiers in Chapters 4 and 5. The terminal characteristics of these devices are related to the effects of fabrication processes, element characteristics, and circuit designs. Typical state-of-the-art specifications of important device parameters are cited, and the practical limitations imposed by those parameters are explored. Specialized techniques for measuring and improving these performance parameters are indicated. The versatility afforded by voltage comparators and operational amplifiers are illustrated through the presentation of useful circuits intended for specific applications. Also included are possible design approaches for modifying the given design to serve additional functional applications.

Chapter 6 presents the operation of several advanced amplifiers that produce specialized functions. Ernest Thibodeaux contributed the material for the first programmable amplifier, and Donald Jones and I prepared the discussion of the PRAM. Robert Webb and Donald Jones cooperated on material preparation for the monolithic, chopper-stabilized amplifier. Don Jones also developed the section in Chapter 6 dealing with the monolithic sample-and-hold amplifier. The unique properties of these advanced amplifiers are presented together with important applications of their utilization.

Chapter 7, which describes analog multiplexers, was developed by Ernest Thibodeaux. Circuit considerations that define the multiplexers' terminal parameters are discussed and illustrated with several practical examples. These applications not only point up the versatility of analog devices but also demonstrate how superior performance often can be achieved through analog switching techniques.

Interfacing between analog and digital devices is covered in Chapter 8. Donald Jones and I cooperated here to describe various digital-to-analog and analog-to-digital conversion techniques and to illustrate useful applications of commercially available converters.

Coverage of a broad spectrum of analog integrated circuits is concluded with the organization and operation of phase-locked loops in Chapter 9. This chapter, developed from a series of application notes prepared by Donald Jones and myself, considers phase-locked loops from the viewpoints of ideal system operation, practical implementation, and selected applications, to emphasize the special characteristics these devices possess.

Although an exhaustive, in-depth treatment of each topic is not provided, this book includes sufficient detail to enable the reader to acquire a working knowledge of the individual topics presented, as well as the language associated with each. A concerted effort has been made to describe the total spectrum of analog ICs by illustrating the interdependence and relationships that exist among our topics. This book focuses not only on *how* analog ICs are fabricated, designed, analyzed, and applied, but also on *why* specific processing technologies, circuit configurations, design philosophies, and so on, are employed to achieve the state-of-the-art device parameters needed for a variety of applications.

I wish to thank all the contributors for their patience and their wholehearted cooperation. It is not possible to enumerate all those whose labors produced the information compiled in this book, but there are several people to whom I am especially indebted and without whose efforts this book surely would not have been completed. First among these is James E. Dykes, whose constant support of the project from inception to completion was the prime factor enabling the various deadlines to be met. Support and many helpful suggestions during the planning and review phases were provided by T. Lamar Clark, Eduardo Fernandez, and Jack Kabell. A special thanks is due Donald Sorchych for his support of the project. Thanks are also due Donald Greene for his counsel during the project. And to Denise Kelly, Debby Finnegan, Deborah Ariail, Kathleen Carroll, Karen Minnis, and Glenys Chafin, my sincerest appreciation for their tireless efforts to put our frequently unintelligible scribbling to a comprehensible form.

To Demetrius T. Paris, Director of the School of Electrical Engineering, Georgia Institute of Technology, and to Georgia Tech, I owe a debt of gratitude for support of this project.

And finally to my wife, Mary Nelle, for her always cheerful support and help with whatever needed to be done, my most sincere appreciation.

For today's technical community, it is sufficient that this work provide a broadened spectrum of analog technology. But for those educators and students who follow, forging tomorrow's technical community, all of us who have participated in the accumulation and preparation of this book ardently hope that the effort will in some small way serve as a stepping stone, one previously unavailable, toward that "state of the art" to come.

J. A. CONNELLY

Atlanta, Georgia
August 1974

CONTENTS

ANALOG INTEGRATED CIRCUITS

CHAPTER 1

INTEGRATED · CIRCUIT FABRICATION

A monolithic integrated circuit is a collection of electrical components formed into and onto a single piece of semiconductor material and interconnected in such a way that a circuit capable of performing a desired function is generated. Individual circuit components usually are formed so that they are electrically isolated from one another. All circuits and processes considered in this book are formed in and on a single silicon substrate, using planar* technology. The term "planar" should not be considered in its literal definition suggesting a plane surface of two dimensions. In the context of integrated circuits, planar refers to the formation of electrical components both on the surface of the substrate and extending below the substrate surface. However, the plane surface of the semiconductor remains relatively flat and unaltered throughout the sequence of fabrication steps employed to form the individual components.

In planar technology, numerous identical circuits are simultaneously formed on a circular silicon disk commonly called a slice or wafer. Figure 1.1a shows a sample wafer, which is typically 2 in. in diameter and 15 mils† thick. A matrix of identical circuit patterns covers the entire wafer surface. Depending on the size of the individual circuit to be integrated, this matrix may contain anywhere from about 100 to more than 1000 circuits on a single wafer (see Figure 1.1b). The various fabrication steps used to form the circuits are performed simultaneously on every member of the matrix as the wafer is processed through each operation in the fabrication sequence. After all fabrication steps have been completed, the

---

* Planar is a patented Fairchild process.
† 1 mil = 10^{-3} in.

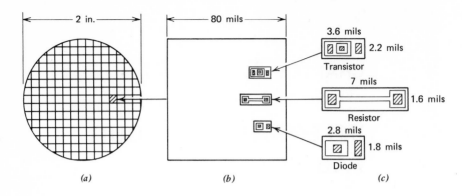

Figure 1.1 Fabrication of an integrated circuit. (*a*) Starting wafer. (*b*) Integrated circuit chip. (*c*) Individual circuit elements. (Adapted from R. G. Hibberd, *Solid State Electronics*, McGraw-Hill, New York, 1968, courtesy of Texas Instruments, Inc.)

wafer is scribed, and the individual integrated circuits (called chips) are recovered for mounting in the appropriate package. Figure 1.2 is a microphotograph of an IC chip. Typically analog integrated circuit chips contain more than 50 individual circuit elements. The individual circuit elements are formed on the silicon wafer in electrically isolated regions.

In a manner that is explained subsequently, selected impurities are introduced into the silicon wafer. These impurities alter the molecular lattice structure of the semiconductor material, causing selected regions on the silicon wafer to possess either an excess or a deficit of electrons in the outer valence shell of the silicon atoms. Excess electrons in a region create N-type semiconductor material; a shortage of electrons produces P-type material. Boundaries where P- and N-type materials abut form P–N junctions of limited surface and vertical extent. The formation and geometries associated with these PN junctions produce the various integrated circuit components of resistors, capacitors, diodes, and transistors. An insulating layer is formed over the surface of these components, and portions are selectively removed to permit access and contact to the different terminals of the components where they intersect the surface. The components are then interconnected by a patterned metal film on the insulator surface in much the same way that components are connected on a single-layer, printed-circuit board. We will now discuss the individual steps basic to the fabrication of all monolithic ICs.

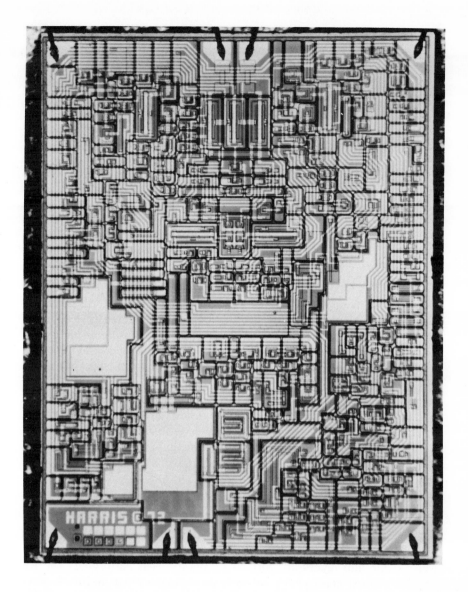

Figure 1.2 Integrated circuit chip. The HA-2900 chopper-stabilized operational amplifier containing over 200 active devices and over 30 passive components.

1.1 BASIC FABRICATION STEPS

The five basic processes, used to fabricate and interconnect all components integrated on a monolithic substrate, are as follows:

1. Oxidation.
2. Photo resist and etching.
3. Diffusion.
4. Epitaxial growth.
5. Thin-film deposition.

A brief discussion of each basic process provides a foundation for later development in this chapter of more specialized fabrication techniques.

Oxidation

In oxidation, the first basic fabrication step, the silicon wafer is heated to a relatively high temperature (800–1200°C) and subjected to oxygen or saturated water vapor. Atoms on the surface of the silicon unite with oxygen, forming a thin layer of amorphous silicon dioxide (SiO_2), which acts as an insulator. When reasonably thin (<50 kÅ*), this SiO_2 layer adheres tightly to the silicon wafer surface. The layer of SiO_2 is nonporous and continuous across the surface of the wafer; as such, the layer acts as an insulator, protecting the inside portion of the wafer from environmental influences. The oxide layer is insoluble and chemically inert in most acids. However, SiO_2 is soluble in hydrofluoric acid, which can be used to remove the oxide from selective areas on the wafer—the next basic step.

Photo Resist

The second basic fabrication step is called photo resist (PR). The extent and the shape of the surface topography of the various P and N regions as well as thin-film layers are formed by photo-resist patterning. The PR process utilizes glass plates called masks, having clear and dark areas formed much like a photographic negative. The dark areas are repeated in a patterned array that covers the entire mask surface. Each pattern in the array corresponds to one individual circuit, and all patterns on an individual mask are identical. Figure 1.3 presents several single mask patterns as well as a pattern array.

The mask pattern is transferred to the wafer through the photo-resist

* 1 Å = 1 angstrom = 10^{-8} m.

process. Pattern transfer is achieved by first coating the surface of the wafer with a thin (~ 7 kÅ) layer of light-sensitive liquid material called photo resist. This chemical material forms a thin transparent film on the wafer surface. Next the pattern on the mask is aligned to the previous pattern (or patterns) on the wafer with an alignment system while the mask and wafer patterns are observed through a microscope. When the required registration is achieved, the wafer is pressed into contact with the mask surface. The mask–photo-resist–wafer sandwich is then exposed to ultraviolet light (UV), which passes through the clear portions of the glass mask and impinges on the photo resist. Dark areas in the mask prohibit the UV light from reaching the photo resist. The UV light that strikes the PR causes it to polymerize. After exposure, the wafers are immersed in a solution of xylene, which dissolves the unpolymerized PR under the dark areas of the mask and leaves the polymerized PR on the wafer. Next the wafer is immersed in an etching solution of hydrofluoric acid, which dissolves the exposed oxide layer where the photo resist was removed. The polymerized PR does not dissolve in the hydrofluoric acid solution. Thus the phoro-resist step is used to produce openings through the SiO_2 layer in selected areas on the wafer as patterned by the mask. These openings provide direct access to the silicon substrate, where impurities are introduced in the next basic step, diffusion. Following etching, the remaining polymerized PR is removed using a solution of sulfuric acid. This acid dissolves only the PR and does not affect the oxide layer. Frequently complex organic acids are used to remove the polymerized photo resist, since these acids do not attack aluminum, which is employed in a subsequent fabrication step to interconnect circuit components.

Diffusion

Diffusion is the third basic process used to form the various P and N regions. The electrical components are produced through careful attention to the placement, geometries, and interconnections of the P and N regions. In the diffusion process, thermally agitated particles having random motion redistribute themselves, moving from regions of high concentration into regions of lower concentration, until equilibrium of concentration is established. The diffusion process can be described mathematically by

$$\frac{\partial N(x, y, z, t)}{\partial t} = D(T) \, \nabla^2 N \tag{1.1}$$

where N is the concentration of the impurity atoms at a spatial point x, y,

6

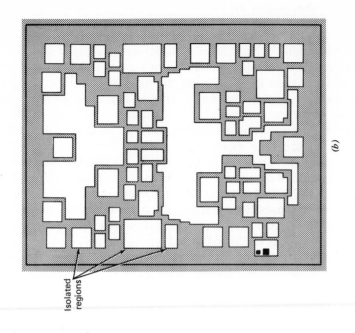

Isolated regions

(b)

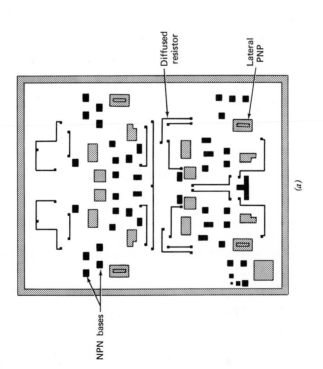

Diffused resistor

Lateral PNP

NPN bases

(a)

Figure 1.3 Sample masks used for photo-resist step to create (a) diffused resistors and NPN bases in JI, (b) isolation islands in DI, (c) NPN bases and diffused resistors in DI, (d) reduction of mask (c) to form pattern array.

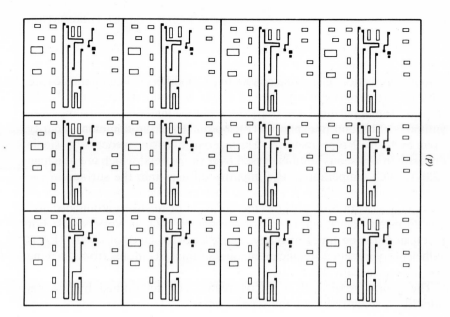

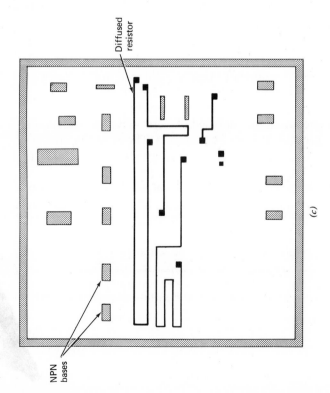

Diffused resistor

NPN bases

(c)

(d)

Figure 1.3 (Cont.)

7

z and at a time t, and D is the diffusion coefficient, which is itself a function of temperature T.

In IC fabrication, elemental conductivity modifiers called impurities are diffused into the crystal structure of silicon to produce the N- and P-type regions. The impurities used to produce N-type material are arsenic (As), antimony (Sb), and phosphorus (P); boron (B) is the common impurity used to form P-type material. During one diffusion process, a single impurity in a gaseous state is passed over the wafers, which are at a high (800–1250°C) temperature. The impurity diffuses into the surface of the wafers wherever silicon is exposed. Surface areas of the wafers previously coated with SiO_2 prohibit the diffusion of impurity atoms because the diffusion coefficient of the impurities in SiO_2 is very small. For practical diffusion times of many hours, an oxide layer of about $7\,k\text{Å}$ is sufficient to prevent a significant number of impurity atoms from reaching the silicon beneath the oxide.

The characteristics of a silicon region are determined by the net impurity concentration of the region, or

$$N_{net} = N_A - N_D \qquad (1.2)$$

where N_A is the acceptor (P-type impurity) concentration and N_D is the donor (N-type impurity concentration). The region is called P-type if N_{net} is positive and N-type if N_{net} is negative.

When impurities of one conductivity type are diffused into a region of opposite impurity type, a P–N junction is formed. Because impurity atoms diffuse from a region of high concentration at the wafer surface to a region of lower concentration, diffused layers have a maximum impurity–atom concentration at the wafer surface. The impurity concentration decreases in a monotonic fashion with distance into the wafer. The P–N junction occurs along a boundary where the concentration of the impurity atoms has reached a concentration level equal to the impurity concentration in the basic material into which the diffusion was made. In terms of Equation 1.2, the P–N junction occurs along a boundary where N_{net} is zero.

The diffused layer is characterized by the important properties of junction depth and sheet resistance. Both properties are characteristics achieved by control of the diffusion process and are independent of the shape of the oxide aperture through which the diffusion is made. The junction depth is the distance measured from the wafer surface to the P–N junction boundary. The sheet resistance is a measure of the effective doping of the layer. The sheet resistance can be expressed as

$$R_s = \frac{\rho_{avg}}{x_j} \qquad (\Omega/\square) \qquad (1 \quad)$$

where ρ_{avg} is the average resistivity of the diffused region in ohm-centimeters and x_j is the junction depth in centimeters. Sheet resistance has the dimensions of ohms, which is also referred to as *ohms* per square ($\Omega/\square$). The average resistivity of the region is given by

$$\frac{1}{\rho_{avg}} = \frac{q}{x_j} \int_0^{x_j} N(x)\mu(x)\, dx \qquad (1.4)$$

where q is electronic charge (1.6×10^{-19} C), N is the impurity concentration (atoms/cm³), and μ is the impurity mobility function (cm²/V-s). Note that both N and μ are functions of the distance from the wafer surface x.

The effective resistance between the two ends A and B of a diffused region of length l, width w, and thickness x_j as shown in Figure 1.4 can be found according to

$$R = \frac{\rho_{avg}}{x_j}\frac{l}{w} = R_s \frac{l}{w} \qquad (\Omega) \qquad (1.5)$$

The ratio of l/w characterizes the pattern defining the topological extent

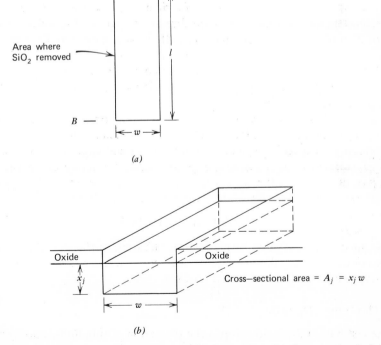

(a)

(b)

Figure 1.4 Geometry of a diffused resistor. (*a*) Top view. (*b*) Cross-sectional view.

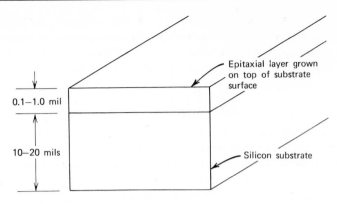

Figure 1.5 Epitaxial growth.

of the diffusion and is called the number of squares of the pattern. Hence the units specified for sheet resistance are ohms per square.

Epitaxial Growth

Epitaxial growth is the fourth basic fabrication process in which a layer of semiconductor material is formed directly on top of the silicon wafers (see Figure 1.5). This epitaxial growth is accomplished by passing a gaseous stream containing two or more compounds over a heated substrate under conditions allowing the compounds to react chemically, decomposing and leaving the desired semiconductor layer on the surface of the heated substrate.

In IC fabrication, epitaxial growth is commonly used to form an N-type, single-crystal, silicon layer on a P-type silicon substrate as part of fabrication of junction-isolated circuits, described later. The basic silicon layer is formed by passing gases of silicon tetrachloride ($SiCl_4$) and hydrogen (H_2) over the wafer, which is heated to about 1150°C. At this high temperature, the H_2 reduces the $SiCl_4$ according to

$$SiCl_4 + 2H_2 \rightarrow 4HCl + Si \qquad (1.6)$$

leaving elemental silicon formed on the substrate surface. A gas of AsH_3, introduced as the silicon is being formed, releases arsenic, causing an N-type doping of the epitaxial surface.

Thin-Film Deposition

The last basic fabrication step is the formation of thin films (approximate thicknesses between 100 Å and 20 kÅ). These films are deposited on the oxide coating of the silicon wafers to interconnect the individual components producing the desired circuit configuration. Thin films are also used to form metal-film resistors having thermal properties different from those of diffused resistors, as we explain further later in this chapter.

Thin films are usually deposited in a high-vacuum environment (absolute pressure $\sim 5 \times 10^{-6}$ mmHg) where the mean free path between the source of the film and the wafer surface is much greater than their separation. A material such as aluminum is heated until it vaporizes and sends off a gaseous stream. The surfaces of wafers facing this stream become coated with the aluminum material. This process for thin-film deposition is called evaporation.

Another thin-film process, called sputtering, is particularly useful for depositing high-melting-point compounds and materials such as molybdenum (Mo). Sputtering is performed in a low vacuum (absolute pressure $\sim 10^{-4}$ mmHg). Argon or another inert material is deliberately introduced into the evaporation system, where it becomes ionized and accelerated toward a target (cathode) of molybdenum. The ionized argon atoms dislodge molybdenum atoms by momentum transfer. Thus Mo atoms are "sputtered" into the surrounding area where the wafers are located, coating the wafer surfaces.

1.2   STANDARD ISOLATION TECHNIQUES

As was discussed in the preceding section, all electrical components in an integrated circuit are formed directly on a monolithic substrate. To avoid interactions among the components, a technique to provide component isolation must be employed. Various approaches for achieving isolation have been developed, including junction isolation (JI), dielectric isolation (DI), air isolation, triple diffusion, and polyplanar technology. In analog circuits, isolation is accomplished more commonly with the JI approach, with DI being the second most common. We now discuss these two important isolation techniques.

Junction Isolation

The most common method for isolating the P and N regions in which the circuit components are built is the junction isolation process, in which each isolation region is surrounded on all sides except the top surface by a reverse-biased P–N junction. Isolation of the top surface is provided through a layer of insulating silicon dioxide.

The process sequence used to form a junction-isolated wafer starts

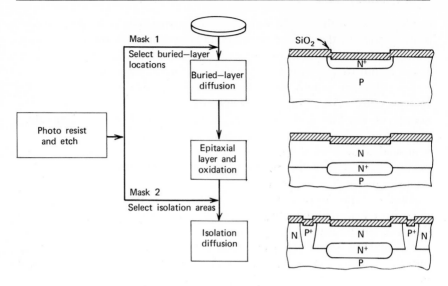

Figure 1.6 Process steps and cross-sectional view of wafer fabrication for junction isolation. (Adapted from *Analysis and Design of Integrated Circuits*, C. S. Meyer, D. K. Lynn, and D. J. Hamilton, eds., McGraw-Hill, New York, 1968, courtesy of Motorola Incorporated.)

with a P-type wafer, as in Figure 1.6. The slice is oxidized, and portions of the oxide are selectively removed from the regions in which islands are to be formed. A low-resistivity, N-type diffusion is made into these regions, and it forms a region called the buried layer. The purpose of creating a buried layer is discussed subsequently. Following the completion of the buried-layer diffusion, all silicon dioxide is removed from the wafer surface and an epitaxial layer of N-type silicon is grown on the surface. This epitaxial layer is then oxidized. Through the photo-resist and masking processes, continuous patterns are etched through the oxide, forming many silicon dioxide islands separated by a continuous pattern of exposed silicon. Below each island is the N+ region formed by the buried-layer diffusion. A high concentration of P-type impurity (isolation diffusion) is diffused into the oxide apertures. These impurities penetrate through the epitaxial growth until they reach the original P substrate, completing the isolation diffusion processes.

Junction isolation produces semiconductor islands bounded on the bottom by the P substrate, on the sides by the P isolation diffusion, and on the top by the silicon dioxide layer. The isolation diffusion and substrate form one continuous region of P-type material. In the finished circuit, this P-region is connected to the most negative voltage available to ensure that it remains reverse-biased with respect to each of the N islands embedded in this P substrate.

Perfect electrical isolation is not achieved by the JI technique. Some parasitic capacitance is associated with the reverse-biased P–N junction. This capacitance is typically about $0.06\,pF/mil^2$ for an analog circuit. Furthermore, some parasitic leakage current is set up at the P–N junction. The magnitude of the leakage current is typically about $1\,pA/mil^2$ for an analog circuit at 25°C. As is true for other leakage currents in planar silicon junctions, this current doubles for approximately every 10°C increase in temperature.

In addition to the passive parasitics of junction capacitance and leakage current, active parasitic devices can occur in the JI process. The substrate–island P–N junction can interact with other P–N junctions in the island to form parasitic PNP devices and PNPN devices (silicon-controlled rectifiers = SCRs). Extreme care must be exercised in device design and circuit topology to ensure that these active parasitic devices do not degrade the intended circuit performance.

Dielectric Isolation

Many of the parasitic problems of junction isolation can be reduced or eliminated by using dielectric isolation. The DI process (Figure 1.7) begins with an N-type substrate into which a high concentration of N-type conductivity modifiers has been diffused to form an N+ buried layer. The silicon surface is then oxidized. Through the photo-resist step, a continuous pattern is formed in the oxide surrounding each region, which becomes an isolated island similar to the isolation patterns formed in the JI process. Exposed silicon is then etched away to a depth greater than the required final island thickness. The slice is reoxidized to form an oxide layer on the surfaces exposed by the silicon etch. Then silicon is epitaxially deposited on this oxide, filling in the etched areas and forming a layer over the entire wafer surface. Because the epitaxial growth is deposited on noncrystalline oxide, a polycrystalline layer results rather than the usual single-crystal layer formed in JI. The DI process is completed when the original wafer is lapped and polished from the side that was not etched. Polishing continues until the lapping plane intersects the etched areas.

In the DI process the sides and bottom of the islands are coated with oxide and embedded in the deposited polycrystalline silicon. These islands are passively isolated by the oxide insulator. Hence no bias voltage need be applied to the substrate to ensure isolation, and no active parasitic devices are created in the DI structure. However, some passive

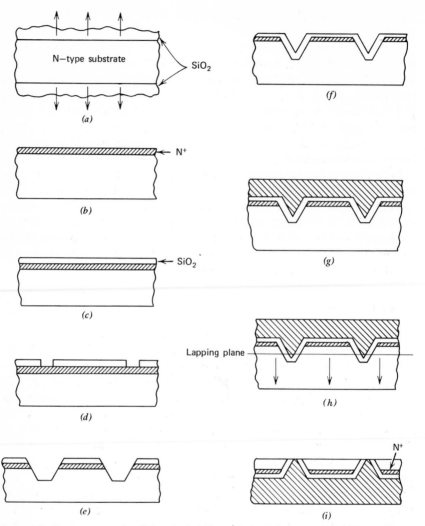

Figure 1.7 Process steps for dielectric isolation. (*a*) Surface preparation, (*b*) N⁺ buried-layer diffusion, (*c*) masking oxide, (*d*) isolation pattern, (*e*) silicon etch, (*f*) dielectric oxide, (*g*) polycrystalline deposition, (*h*) backlap and polish, (*i*) finished slice.

parasitic capacitance of about $0.02 \, pF/mil^2$ is associated with the isolation. Also, some leakage current is set up between the island and the substrate through the oxide isolation. This current is typically 1 pA at 25°C and 10 pA at the extreme temperature of 150°C. Since reverse-biased PN junctions are not used for isolation in DI, parasitic leakage current does not double for each 10°C temperature rise as with JI.

1.3 FORMATION OF MONOLITHIC CIRCUIT COMPONENTS

Circuit components are structured into the individual islands created through the JI and DI techniques. The basic process used to form circuit components employs two additional diffusions and one thin-film evaporation.

The first diffusion process starts by oxidizing the isolated wafer surface, as illustrated in Figure 1.8. The oxide is patterned through the PR process, and a P-type diffusion of typically 100 $\Omega/\square$ and 3 μm* junction depth is made into the SiO_2 apertures. The wafer is oxidized again, and the oxide is patterned a second time by photo-resist processing. A high-concentration N-type (N+) phosphorus diffusion of 2 $\Omega/\square$ and 2.5 μm junction depth is made into these apertures. Oxide is again formed over this diffusion, and another PR patterning step is performed to remove oxide from areas of each terminal of each component. Aluminum

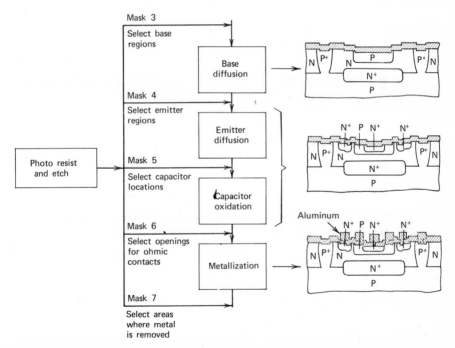

Figure 1.8 Basic component fabrication processes for JI. (Adapted from *Analysis and Design of Integrated Circuits*, C. S. Meyer, D. K. Lynn, and D. J. Hamilton, eds., McGraw-Hill, New York, 1968, courtesy of Motorola Incorporated.)

* 1 μm = 1 micron = 10^{-6} m = 0.03937 mil.

is evaporated over the entire slice, using the thin-film process. The aluminum contacts the silicon wafer only in areas where oxide had been removed previously by the PR process. The aluminum is patterned by a PR operation to interconnect the components in the desired circuit configuration.

The first of the two additional diffusions produces a P–N junction between the original N-type island and the P-type conductivity modifier. This P–N junction will eventually become the base–collector junction of the NPN transistor being formed. This P diffusion is also used to form resistors by making some of the diffusions in long, narrow apertures and contacting them at each end (see Figures 1.3 and 1.4). Any number of resistors can be formed in the same N-type island and isolated from one another by connecting the N island to the most positive voltage in the circuit. This ensures that every P-resistor–N-island junction is reverse-biased. Thus junction isolation among the resistors is achieved in a similar method as the N islands in the P substrate are isolated.

The second diffusion used to produce circuit components is an N diffusion that forms the emitters of the NPN devices. A high doping concentration is used, making this diffusion produce N+ material. Heavily doped N+ regions are also simultaneously formed in the N island for interconnecting the aluminum metal to the transistor's collector region. This process yields an integrated, double-diffused, NPN transistor, whose typical dimensions (see Figure 1.9) indicates that a relatively wide spacing exists between the emitter and the collector contacts. The N-type

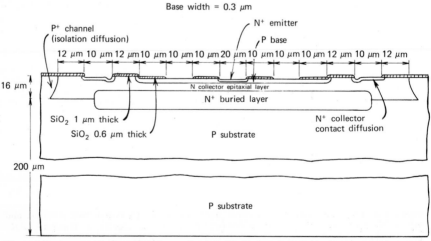

Figure 1.9 Junction-isolated NPN transistor, indicating typical dimensions. (Adapted from *Analysis and Design of Integrated Circuits*, C. S. Meyer, D. K. Lynn, and D. J. Hamilton, eds., McGraw-Hill, New York, 1968, courtesy of Motorola Incorporated.)

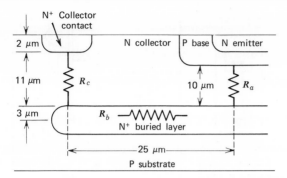

Figure 1.10 Cross-sectional view of an NPN transistor, illustrating function of the buried layer. (From *Electronic Integrated Circuits and Systems* by F. Fitchen, © 1970 by Litton Educational Publishing, Inc. Reprinted by permission of Van Nostrand Reinhold Company.)

collector region, being lightly doped, presents a relatively high resistance to charge flow between the emitter and the collector contact. By fabricating a buried layer of highly doped material directly beneath the collector region, the emitter–collector resistance is reduced considerably. Figure 1.10 gives an expanded view of this buried-layer region. The emitter electrons cross the base region, entering the N-collector region, where they flow vertically through an effective resistance R_a to the buried layer. The electrons then move through the effective buried-layer resistance R_b, which is much smaller than the resistance of the collector region. Finally the electrons move through R_c to the collector contact, where they exit the device.

Figure 1.11 depicts the various components in a typical vertical structure formed in the JI islands during the normal processing steps. In addition to the normal NPN transistor and diffused resistor, the figure shows profile views of lateral and substrate PNP transistors and a metal-oxide semiconductor (MOS) capacitor. We now discuss the variations of the basic JI process used to produce these components.

Lateral PNP Transistors

A lateral PNP transistor can be formed at the same time the base diffusion is made for a normal NPN transistor. The transistor is called lateral because charge flow in the base of the device is parallel to the surface of the substrate. An isolated N island is selected, and P-type impurity is diffused into two adjacent, closely spaced regions. The N island acts as the transistor base, with one P-diffused region becoming the emitter and the other P diffusion becoming the collector.

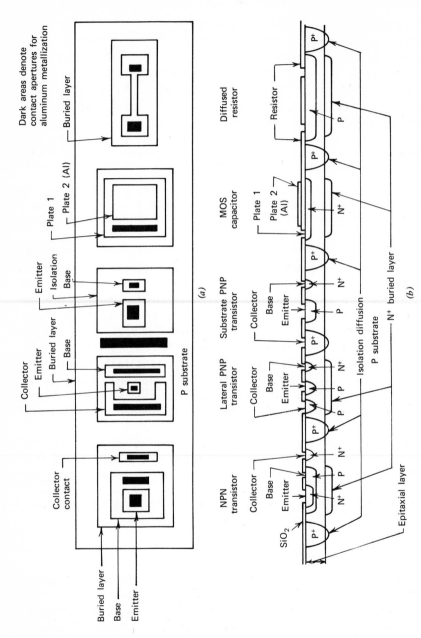

Figure 1.11 Structures of various components formed in the junction-isolation process. (*a*) Topological view. (*b*) Cross-sectional view.

18

Because of mask fabrication and chemical etching tolerances associated with the photo-resist process which determine the base width (emitter-to-collector distance), the lateral PNP transistor is a low-performance device with a typical β* between 5 and 50 and a typical f_T† of only a few megahertz. Furthermore, lateral PNP transistors are not capable of efficiently carrying currents in excess of about 1 mA. Despite these practical limitations, lateral PNPs are widely employed in analog circuit designs because they can be formed without requiring any additional diffusion steps beyond those necessary to create NPN transistors. Hence lateral PNPs are a "free" device in the sense that they can be made with no increase in the cost of processing a wafer over that required to make NPN transistors alone.

Substrate PNP Transistors

Another type of PNP transistor is called the substrate PNP. This device is formed exclusively in junction-isolated circuits and uses the P substrate as the collector, as in Figure 1.11. The N island constitutes the base, with a P diffusion being the emitter. The P diffusion to form the emitter of the substrate PNP transistor is the base diffusion of the conventional NPN devices. Substrate PNPs have somewhat higher β's (~ 100) and larger f_T's (~ 10 MHz) than lateral PNPs and can handle currents up to about 5 mA for a reasonable size device. However, extensive usage in circuit designs is limited because the collector (which is the wafer substrate) must be connected to the most negative voltage in the circuit.

MOS Capacitors

The N+ diffusion used in producing the emitter and collector contacts for NPN transistors can also be used to form the "lower" plate of a parallel-plate capacitor of the type previously shown in Figure 1.11. This capacitor is called an MOS capacitor because the aluminum *metallization* acts as the top plate and the SiO_2 is the *oxide* that acts as the dielectric material grown over the N+ *semiconductor* region. The capacitance value is given by

$$C = \frac{A\varepsilon}{t} \tag{1.7}$$

where A is the plate area of either the top or bottom plate (whichever is smaller), ε is the dielectric constant of SiO_2, and t is the oxide thickness. Typical thicknesses of MOS capacitors range from about 800 Å to 4 kÅ.

---

* The symbol β, frequently called h_{fe}, denotes the short-circuit current gain of the transistor in the common-emitter configuration.
† The frequency at which the magnitude of β extrapolares to unity is denoted f_T.

Junction Capacitors

A reverse-biased P–N junction has an associated intrinsic depletion-layer capacitance that can be utilized as a circuit component in an IC. The value of this capacitance is given by

$$C \approx A_j \left(\frac{q \varepsilon N}{2V} \right)^{1/2} \tag{1.8}$$

where A_j is the junction area, N is the impurity concentration of the lightly doped side of the junction (N island in this case), and V is the voltage applied which reverse-biases the junction. As Equation 1.8 shows, the size of the capacitance depends on the applied voltage. Furthermore, the junction capacitor is polarized because the P side of the junction must always be at a voltage more negative than the N side, to ensure that the junction remains reverse-biased. Under forward-biased conditions, the intrinsic diffusion capacitance produces the dominant effect and causes a low impedance between the capacitor terminals. Typically, the junction capacitor has an associated leakage current of 1 pA/mil^2 at 25°C, which doubles for each 10°C temperature rise.

The basic processes and technologies we have outlined can be used to build both analog and digital integrated circuits. However, since the applications of analog and digital circuits are altogether different, specific ways of structuring the basic components vary in accordance with intended circuit function. For example, analog circuits require matched pairs of transistors having relatively high β's and large f_T's. Analog transistors also must be built to achieve high reverse-breakdown voltages and low leakage currents.

To achieve high breakdown voltages, N-type islands having a high resistivity (about 5 Ω-cm for a 50 V BV_{CEO} NPN) and a large base-buried layer separation (0.4 mil) are used. To achieve high current gains and low leakage currents, we must observe special fabrication precautions to prevent contamination of the wafer with undesirable impurities. Close device matching is accomplished most directly through careful control of component dimensions. We now turn to several special techniques used to achieve analog devices with particular characteristics.

1.4   SPECIAL DEVICE PROCESSES

Super-β Transistors

In many analog applications input transistors are required to have very small base currents (typically < 10 nA). Such low base current levels can be achieved only at the sacrifice of other device parameters such as reverse-breakdown voltage. Super-β devices are designed to maximize β at the expense of reduced-breakdown voltages of BV_{CEO} and BV_{CBO}. Super-β devices have a typical BV_{CBO} breakdown voltage between 10 and 20 V and a BV_{CEO} breakdown between 5 and 10 V.

A useful expression developed by Warner and Fordemwalt (18) can be adapted to relate $1/\beta$ to the physical parameters of the device as

$$\frac{1}{\beta} \approx \frac{\rho_E X_B}{\rho_B X_E} + \frac{X_B{}^2}{4L_{nB}} \tag{1.9}$$

where ρ_E = emitter resistivity
ρ_B = active base-region resistivity
X_B = base width
X_E = emitter-junction depth
L_{nB} = diffusion length of minority carriers in base region

The appearance of X_B in the numerator of both terms of Equation 1.9 shows that the base width must be as small as possible to maximize β. Reducing the base width improves the emitter efficiency by reducing the total number of impurity atoms in the active-base region and also lessens the percentage of minority carriers that recombine as they move across the base region.

The standard analog processes for transistor fabrication can be altered in one of two ways to form super-β devices. One technique modifies the standard process by forming an extra P-base diffusion which is slightly shallower than the normal NPN transistor base. The normal N+ diffusion is then used to form emitters in the bases of all transistors. This technique produces normal NPNs in the standard P-base diffusion and super-β NPNs with narrow base widths because of the shallower base diffusion.

The second technique for super-β processing uses the standard P-base diffusion followed by two separate N+ emitter diffusions. The first N+ diffusion is made slightly deeper into the P material than the standard depth. Deeper emitter penetration reduces the base width, thereby creating the super-β device. The second N+ diffusion is not made into the super-β transistors but is utilized only for the regular NPN transistors.

Junction Field-Effect Transistors (JFET)

When input impedances larger than those achieved with super-β devices are required, field effect transistors (FETs) are commonly used. These devices have a channel for conducting current between two contacts termed the source and drain. The channel conductance can be varied by applying a control voltage to a third terminal called the gate.

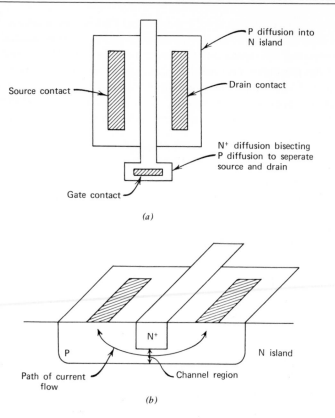

Figure 1.12 P-channel JFET showing (*a*) topology and (*b*) cross-sectional view.

The first type of planar FET to be successfully produced was the junction FET or JFET. Topological and cross-sectional views for a JFET appear in Figure 1.12. A P-type channel region is bounded on all sides by an N region forming a P–N junction that is normally reverse-biased. A second P–N junction is formed between the P channel and the N+ gate region. This junction is also reverse-biased, and a depletion region is set up directly beneath the N+ gate. The width of this depletion region is controlled by the gate-to-source biasing voltage. Increases in the biasing voltage widen the depletion region, causing it to spread outward from the P–N junction, thereby reducing the cross-sectional area of the channel and constraining the drain current.

In principle, the JFET structure could be formed using the standard P-base diffusion to form the P region of the FET. The N+ diffusion could be used to form the upper gate, which defines the channel region.

However, JFET devices are not made in this manner because overly thick channel regions are produced. A variation of the gate-to-source voltage modulates only the edges of the channel without affecting its center region. The channel region must be narrowed to achieve a JFET device having useful characteristics. This can be accomplished by using either of the methods used to form super-β devices. In fact, both super-β and JFET devices may be made simultaneously with the same process steps.

Metal-Oxide Semiconductor Field Effect Transistors (MOS FET)

The second type of FET used frequently in analog integrated circuits is the MOS device in Figure 1.13. The device has a gate region that is insulated from the body of the transistor by a dielectric substance. The gate is formed during the interconnect metallization step and lies on a thin (typically 1 kÅ) layer of SiO_2 which acts as the dielectric. The silicon surface beneath the gate oxide is the body of the device. P–N junctions on either side of the gate form the source and drain.

A P-channel MOS FET (which has an N-type body) may be formed by a simple modification of the standard transistor process. The P-base diffusion is used to form source and drain regions. The gate region is formed after the N+ diffusion by etching away all oxide lying on the transistor body between source and drain. The slice is reoxidized to achieve the desired gate–oxide thickness. The wafers are then processed to completion using the standard process steps; the metallization step produces the gate of the device.

Parasitic PMOS devices occur in the standard process whenever an interconnect metallization line crosses two P diffusions with an N surface between them (i.e., where Al crosses two P resistors in the same island or where a metallization line crosses a P base and P-isolation diffusion in an N-collector island). In this parasitic structure, the metallization acts as the gate, the more negative of the two P diffusions becoming the drain and the other P diffusion becoming the source. The N surface is the body of the PMOS device. Typically such parasitic devices have relatively large threshold voltages between -20 and -30 V. However, if the gate-to-source voltage should extend into this range, conduction will occur from drain to source, detracting from the desired circuit operation.

N-channel MOS devices can be formed during the standard process by adding a P-type body diffusion before the regular P-base diffusion. This body diffusion forms a P-surface layer having a surface concentration suitable for formation of N-channel devices possessing the desired threshold voltage. The surface concentration of P-type impurities associated with the body diffusion is approximately 100 times less than that

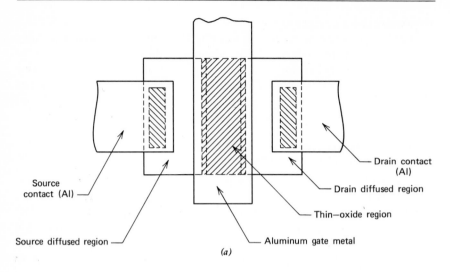

Source contact (Al)

Source diffused region

Drain contact (Al)

Drain diffused region

Thin—oxide region

Aluminum gate metal

(a)

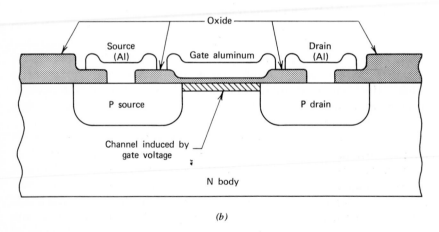

(b)

Figure 1.13 MOS FET structure showing (a) topological view and (b) cross-sectional view of a P-channel device.

of a typical P-base diffusion. Consequently, the P-base diffusion cannot be used to form the NMOS body, and an additional body diffusion must be employed. Subsequent processing steps are identical to the PMOS process, and the emitter diffusion is used to form the NMOS source and drain regions. This process yields both NMOS and PMOS devices which can be used to form complementary MOS (CMOS) circuit functions.

High-Frequency Transistors

Many analog integrated circuit applications require good device performance at high frequencies well into the megahertz range. To achieve such performance requires bipolar transistors with exceptionally high values of f_T. Whereas the basic fabrication process yields NPN devices having reasonably large f_T values, PNP devices that possess good high-frequency properties are not obtained unless the basic processes are modified.

Warner and Fordemwalt (18) provide a useful equation for relating f_T to the physical and electrical properties of NPN and PNP devices:

$$\frac{1}{f_T} = 2\pi \times 1.4\left(r_e C_e + \frac{X_B^2}{5D_{nB}} + R_{cs}C_c\right) \qquad (1.10)$$

where C_e = emitter–junction capacitance
$\quad\ C_c$ = collector–junction capacitance
$\quad\ r_e$ = small-signal emitter–junction resistance
$\quad\ R_{cs}$ = series resistance of the bulk-collector region
$\quad\ X_B$ = base width
$\quad\ D_{nB}$ = diffusion coefficient of minority carriers in the base region

The second term of Equation 1.10 involving the base width, X_B, usually has the dominant effect in low and moderate f_T devices. High-frequency devices have very shallow bases to minimize this term without losing high-breakdown-voltage capability.

A high-frequency PNP device formed in the basic DI process is structured with a vertical, double-diffused structure similar to the NPN. A double-diffused PNP utilizes a P-type island as its collector. Therefore, it requires a starting wafer that has isolated P islands for PNP collectors and N islands for NPN collectors, as in Figure 1.14. This wafer is produced through a modification of the regular DI process. Where P-type islands are required, the N silicon is selectively etched away to a depth greater than the final island thickness. The holes are refilled by a selective P-type epitaxial growth. After formation of the two types of buried layer, the wafer is processed through the same steps as the standard DI wafer. During the final polishing step, the polishing plane intersects the P-collector epitaxial regions. The finished wafer (Figure 1.14) has the desired P- and N-type islands.

High-frequency PNP devices are built on the specially prepared wafer using a sequence of four PR-diffusion steps. The first is an N-base PR step with an N-type diffusion to form the base regions of the PNP devices. Next a PR and P-type diffusion sequence is made to form NPN bases and circuit resistors. Third, a PR and P+ diffusion sequence is done

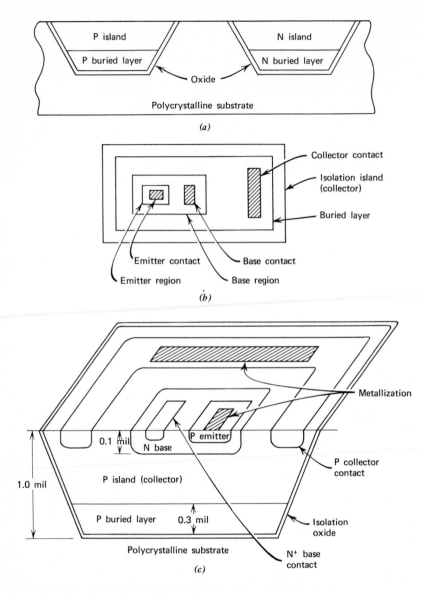

Figure 1.14 The high-frequency process. (*a*) Cross-sectional view of P and N islands for PNP and NPN transistors. (*b*) Topological view showing relative placement of transistor regions. (*c*) Cross-sectional view of high-frequency PNP device formation in the DI process.

to form PNP emitter regions and collector-region contacts. Finally, a PR and N+ diffusion forms NPN emitters and collector regions. Interconnection metallization is formed just as in the standard process. Figure 1.14c is a cross-sectional view of a finished high-frequency PNP transistor in the DI process.

As with the standard process, the high-frequency process can be modified by the addition of an MOS gate–oxide sequence. This modification yields PMOS devices in N islands and NMOS devices in the P islands. The PMOS device characteristics are similar to those of the modified standard process. However, the NMOS devices have a lower threshold voltage because the surface doping of the P island in which they are built is lower than is usually achieved through the "front-side" diffusion process. The threshold voltage can be made low enough to permit NMOS devices to operate in the depletion mode. Depletion-mode MOS FETs are very useful as analog amplifying devices, particularly in high-frequency circuits.

Special Resistor Fabrication Processes

Metal-film Resistors   We have thus far considered the thin-film process only to form an interconnection metallization. Another important application of metal-film technology is the formation of metal-film resistors on top of the wafer-oxide surface. These resistors can be made by depositing or sputtering materials such as nickel chromium (NiCr), molybdenum silicon (Mo_2Si), and chromium silicon (CrSi). Metal-film resistors have better temperature coefficients (< 100 ppm/°C) than diffused resistors, as well as better matching between components.

The basic process used to form metal-film resistors is the same for all resistor materials. A thin film of the resistor material is evaporated or sputtered onto the wafer surface after contact apertures have been etched through the passivating oxide to the device contact regions. The film is then patterned to form the resistor geometries, using the usual PR and etching sequence. The wafers are cleaned and coated with the metallization thin film and processed to completion in the usual way.

Pinched Resistors   The pinched-resistor fabrication approach is useful for producing relatively large-valued resistors without requiring excessive chip area. However pinched resistors must be employed in circuit positions where performance characteristics are not critical. For example, pinched resistors have larger temperature coefficients (typically 10^4 ppm/°C) than regularly diffused resistors (typically 2×10^3 ppm/°C). Furthermore, the tolerance control of the absolute value of pinched resistors is very poor, being typically between -50 to $+100\%$.

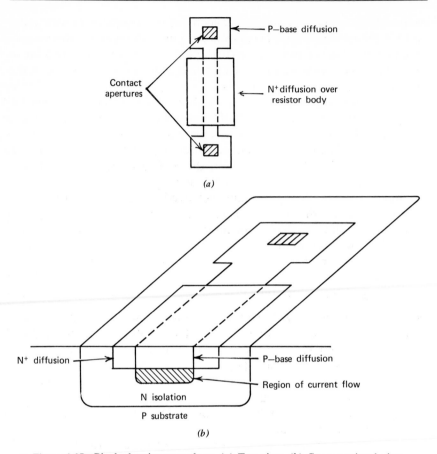

Figure 1.15 Pinched-resistor topology. (*a*) Top view. (*b*) Cross-sectional view.

The structure of a pinched resistor is presented in Figure 1.15. The beginning fabrication step of the pinched resistor is the same P-base diffusion as is used for a normal resistor. However, the second-step diffusion is an N+ emitter diffusion which overlaps the area that would normally have become the resistor body. This overlapping diffusion forms a P–N junction that considerably reduces the cross-sectional area of the channel and constrains the current flow to the lightly doped bottom portion of the original P diffusion. This constraint of the resistor's cross-sectional area greatly increases the sheet resistance of the diffused region from approximately 100 $\Omega/\square$ to about 5 k$\Omega/\square$. Hence larger-valued resistors are formed without requiring excessive chip area.

Ion Implantation Ion implantation is a well-established technology based on the mass spectrometer principle, which has recently been applied to semiconductor fabrication of P–N junctions. The average resistivity tolerances achieved through ion implantation are superior to those due to the basic diffusion processes. With ion implant, ions of the desired impurity atom are created in an evacuated chamber, accelerated by an electric field, and deflected by a transverse magnetic field. The desired impurity ions impinge on the silicon wafer, which acts as a target. The ions possess enough energy to penetrate average distances into the wafer of several hundred to several thousand angstroms. The penetration depth is controlled by the applied electric-field strength.

The implantation is very closely controlled by electronically integrating the ion-beam current and terminating the ion process when the desired ion dose has been achieved. The junction depth is controlled through the dose intensity and the accelerating voltage.

Selective areas on the wafer are masked by photo resist or aluminum thin-film which block the ion stream to achieve the desired wafer pattern. After implantation, the wafer is annealed at a moderate temperature between 400 to 1000°C to electrically activate the implanted ions and to remove crystal damage caused by implanting. Perhaps the most obvious utility of ion implantation for analog circuits is in the fabrication of resistors for ladder networks and other precision applications. Use of this technique can significantly improve resistor matching, as well as control the absolute value of the resistors.

Radiationed-Hardened Devices

Integrated circuits are sometimes subjected to radiation environments— for example, certain applications require them to serve as components in satellites and nuclear weapons systems. ICs are particularly sensitive to two major types of radiation: transient ionizing radiation produced by gamma and X-rays, and neutron radiation, which causes permanent damage by displacing the crystal lattice structure. Each type of radiation produces a different effect on the semiconductor device and circuit performance. A detailed description of these radiation effects can be found in Ricketts (14).

Ionizing radiation is a transient phenomenon that produces an excess of electron-hole current carriers in silicon. Under the influence of gamma rays, circuit operation can be degraded by photocurrent generation, latch-up, or alteration of the surface properties of the semiconductor chip. Of these three effects of ionizing radiation, photocurrent generation is the most important.

Photocurrent generation is the creation due to the ionizing radiation of current carriers in reverse-biased P–N junctions. The peak value of the photocurrent produced is proportional to the junction area and to the radiation intensity. The collector–isolation and the collector–base junctions in conventional transistors are particularly sensitive to photocurrent production because these junctions constitute most of the reverse-biased areas in monolithic devices.

In JI circuits, ionizing radiation can cause the collector–substrate junction to conduct sufficient current to create device latch-up. This occurs when sufficient current exists in a four-layer p-n-p-n structure composed of the emitter–base, base–collector, and collector–substrate junctions in a conventional NPN transistor having an associated PNP parasitic transistor. If this PNP transistor turns on because of ionizing radiation, the p-n-p-n structure will latch up in the conducting mode. Even though photocurrent generation is a transient effect and will temporarily disrupt circuit operation, permanent destruction of the device may occur if the device current is not limited. Latch-up can be eliminated by using DI instead of the conventional JI process.

MOS and JFET devices are more sensitive to ionizing radiation than bipolar transistors because all P–N junctions in the former devices are reverse-biased. Furthermore, the silicon and SiO_2 surfaces in MOS devices are very sensitive to ionizing radiation, which can cause alternations in the threshold voltage. For these reasons, utilization of MOS FET devices in radiation-hardened circuits is generally avoided.

The second important type of radiation for ICs is neutron radiation, where collisions occur between neutrons from the radiation source and the silicon atoms in the IC semiconductor. These collisions permanently displace some of the silicon atoms from their original lattice structure, giving rise to what is called "neutron-displacement" damage. The silicon atoms that have been displaced act as recombination and scattering centers, decreasing the minority-carrier lifetime, τ and the carrier mobility μ. Many transistor parameters such as β, reverse-leakage current, and storage time depend directly on the minority-carrier lifetime. A decrease in the carrier mobility will increase the resistivity of all semiconductor regions. The increase in turn will produce the most harmful effects in high-resistivity regions, such as the collectors of NPN transistors and large-valued resistors. Hence overall circuit performance can be seriously degraded by neutron radiation.

Certain modifications to the basic fabrication process are employed to increase a monolithic circuit's resistance to radiation effects. Although it is difficult to generalize all the techniques used to produce "hardened" devices, some guidelines can be established if we realize that the guidelines themselves are often altered depending on the particular circuit function and the anticipated radiation. To reduce the effect of ionizing radiation, dielectric isolation is generally used to avoid reverse-biased P–N junctions. Furthermore, base–collector junction areas in bipolar transistors are minimized to reduce the effects of photocurrents.

Photocurrents can be minimized by minimizing lifetime. This can be accomplished by gold doping, a technique also used to reduce lifetime in switching devices to minimize storage time.

Thin-film resistors are preferred over diffused resistors to eliminate reverse-biased junction effects. Protective resistors are used in each circuit branch between the power supplies and dc ground to limit transient photocurrents to safe levels.

Neutron radiation also reduces lifetime. The major effect of this radiation is a lowering of h_{FE}* due to reduced base transport factor. Transistors in hardened circuits are made with shallow bases and narrow base widths to minimize the sensitivity of h_{FE} to base transport factor. Also, the initial value of transistors' h_{FE} is made very large to allow for reductions due to neutron radiation.

1.5 CONCLUSIONS

In this introductory chapter, we have discussed the five basic fabrication steps: oxidation, photo resist, diffusion, epitaxial growth, and thin-film deposition. Using these basic techniques, we have illustrated how junction and dielectrically isolated islands can be prepared on silicon substrates on which integrated transistors, resistors, and capacitors of various types are formed. Some necessary modifications to the basic fabrication processes were discussed to demonstrate how devices with special characteristics are produced. Some devices discussed include super-β transistors, JEFTs, MOS FETs, high-frequency transistors, metal-film and pinched resistors, and radiation-hardened devices. Throughout the remainder of this work, and especially in the next two chapters, we make frequent references to these fabrication techniques in explaining the production of analog integrated circuits having special characteristics for diverse applications.

BIBLIOGRAPHY

1. Ahmed, H., and P. J. Spreadbury, *Electronics for Engineers, An Introduction*, Cambridge University Press, London, 1973.

* The symbol h_{FE} denotes the dc, common-emitter, short-circuit current gain.

2. Angelo, E. James, Jr., *Electronics: BJTs, FETs, and Microcircuits*, McGraw-Hill, New York, 1969.

3. Fitchen, Franklin C., *Electronic Integrated Circuits and Systems*, Van Nostrand-Reinhold, New York, 1970.

4. Gray, Paul E., and Campell L. Searle, *Electronic Principles: Physics, Models, and Circuits*, Wiley, New York, 1969.

5. Grebene, Alan B., *Analog Integrated Circuit Design*, Van Nostrand-Reinhold, New York, 1972.

6. Hunter, Lloyd P., *Handbook of Semiconductor Electronics*, McGraw-Hill, New York, 1970.

7. Meyer, Charles S., David K. Lynn, and Douglas J. Hamilton, *Analysis and Design of Integrated Circuits*, McGraw-Hill, New York, 1968.

8. Millman, Jacob, and Christos C. Halkias, *Electronics Devices and Circuits*, McGraw-Hill, New York, 1967.

9. Millman, J., and C. C. Halkias, *Integrated Electronics: Analog and Digital Circuits and Systems*, McGraw-Hill, New York, 1972.

10. Pierce, J. F., and T. J. Paulus, *Applied Electronics*, Merrill, Columbus, Ohio, 1972.

11. Pierce, J. F., *Semiconductor Junction Devices*, Merrill, Columbus, Ohio, 1967.

12. RCA Inc., "Linear Integrated Circuits," RCA Corp., Technical Series IC-42, 1970.

13. RCA Inc., "Linear Integrated Circuits and MOS Devices," RCA Corp. No. SSD-202, 1972.

14. Ricketts, L. W., *Fundamentals of Nuclear Hardening of Electronic Equipment*, Wiley, New York, 1972.

15. Ryder, John D., *Electronic Fundamentals and Applications*, Prentice-Hall, Englewood Cliffs, N.J., 1970.

16. Schilling, Donald L., and Charles Belove, *Electronic Circuits: Discrete and Integrated*, McGraw-Hill, New York, 1968.

17. Schwartz, Seymour, *Integrated Circuit Technology*, McGraw-Hill, New York, 1967.

18. Warner, Raymond M., Jr., and James N. Fordemwalt, *Integrated Circuits: Design Principles and Fabrication*, McGraw-Hill, New York, 1965.

CHAPTER 2

CHARACTERIZING ANALOG INTEGRATED CIRCUIT DEVICES

Effective utilization of monolithic devices requires specialized techniques and models to characterize their performance. In this chapter we treat a number of useful models that have given accurate representations of the behavior of integrated components.

Our approach in presenting these component models is to provide an intuitive relationship between portions of the characterizing model and the fabrication effects discussed in Chapter 1. Our intention is not to develop extremely sophisticated models which account for numerous device influences. Instead we want to promote an intuitive understanding of the relations between actual devices and the characterizing models.

2.1 SPECIAL CONSIDERATIONS OF ANALOG INTEGRATED CIRCUIT MODELS

Under normal operation, most components in analog integrated circuits (AICs) are subjected to small-signal perturbations that slightly modulate the normal current and voltage levels associated with a particular component. Analog circuits differ in this respect from digital circuits, which are subjected to large-signal changes of their component's operating point. Consequently, analog circuit design and analysis requires small-signal component models.

Bipolar transistors in analog circuits are usually modeled for operation in their active region where the base–collector P–N junction is reverse-biased and the base–emitter junction is forward-biased. FETs are normally operated in the "on" condition, where the applied gate voltage exceeds the threshold or pinchoff voltage. Integrated resistors and capacitors in AICs carry some nominal current and have some nominal voltage applied which properly biases these elements for small-signal influences. All the models presented in this chapter are applicable for describing device behavior under small-signal conditions.

2.2 ANALOG INTEGRATED CIRCUIT RESISTOR MODEL

Diffused Resistors

A diffused resistor is a distributed element whose resistance depends directly on its l/w ratio, as indicated in Equation 1.5. This distributed nature requires a model composed of a series of identical segments as shown in Figure 2.1. In this model, the total diffused resistance between the end terminals A and B is R, which is equal to the sum of the individual ΔR segments.

The ΔC capacitance in the model represents the distributed capacitance associated with the reverse-biased P–N junction bounding the resistor diffusion. The value of ΔC can be found using Equation 1.8 by substituting the junction area of a resistor segment, ΔA_j, for the total junction area A_j.

In the diffused-resistor model, the segmented current sources Δi_l account for the leakage currents flowing across the reverse-biased P–N junction. An individual Δi_l segment has an approximate value given by

$$\Delta i_l \approx 1 \ \text{pA} \times \Delta A_j \qquad (2.1)$$

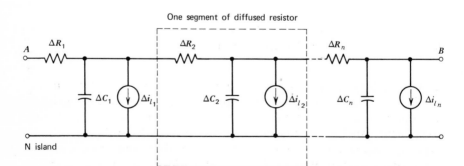

Figure 2.1 Distributed model for diffused resistor. Typical values: $\Delta l = 5$ mils, $\Delta R = 1 \ \text{k}\Omega$, $\Delta C = 0.25 \ \text{pF}$, $\Delta i_l = 3.5 \ \text{pA}$.

where ΔA_j is the junction area of a segment of length Δl. The area of a segment can be found from

$$\Delta A_j = (2x_j + w)\, \Delta l \tag{2.2}$$

Because Δi_l is a junction-leakage current, the value found from Equation 2.1 will double for each 10°C temperature rise.

The choice of the number of segments to be used in the distributed model depends on the signal frequency and the required accuracy. Since the total element should act as a resistor, a basic rule of thumb is to choose the segments small enough so that the impedance of the shunt capacitance is at least an order of magnitude larger than that of the resistor segment, or mathematically,

$$|\Delta Z_c| = \frac{1}{\omega\, \Delta C} \geq 10\, \Delta R \tag{2.3}$$

or

$$\frac{1}{(\omega C / A_j)(2x_j + w)\, \Delta l} \geq \frac{10 R_s\, \Delta l}{w} \tag{2.4}$$

At low frequencies the resistor can be treated as a single lumped-element model containing the resistor R and the leakage current i_l. The leakage current becomes more significant at high temperatures, where it is greatest.

As with all general component models, different applications tend to emphasize different parts of the model. With careful consideration to the model application, simplifications are often possible to reduce the model's complexity. In a precision dc application, for example, we need consider only the leakage current of a diffused resistor, which can be as large as 10 nA at high temperatures. If the nominal current is 1 μA, a large 1% error results. Errors of such a magnitude limit the usefulness of diffused resistors at low current levels.

Diffused resistors have rather large temperature coefficients which vary depending on the doping concentration. A typical variation of the total resistance with temperature appears in Figure 2.2. (For comparative purposes, the thermal effects of a thin-film resistor are also shown.) The temperature coefficient (TC) for the typical diffused resistor is between 10^3 and 2×10^3 ppm/°C. As the curve illustrates, the sign of the TC is positive for temperatures greater than 0°C and negative for temperatures below 0°C. The TC of diffused resistors is controlled by the hole mobility, which has a negative TC, and the carrier concentration, which has a positive TC. The relative contributions of these two factors determine the sign of the overall TC for the diffused resistor.

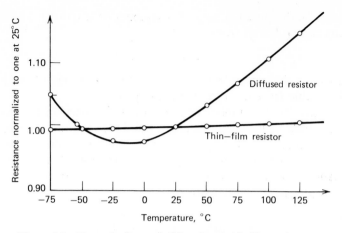

Figure 2.2 Thermal effects of diffused and thin-film resistors.

Thin-Film Resistors

Thin-film resistors can be modeled by a simple RC segmented network (see Figure 2.3). This model is simpler for thin-film resistors than for diffused resistors because it does not contain a current source to represent leakage current. The leakage current through the oxide is so small that it can be neglected. The segmented resistor and capacitor values can be found from

$$\Delta R = R_s\left(\frac{\Delta l}{w}\right) \tag{2.5}$$

and

$$\Delta C = \frac{w\, \Delta l \varepsilon}{t} \tag{2.6}$$

where ε is the dielectric constant of SiO_2 and t is the thickness of the SiO_2

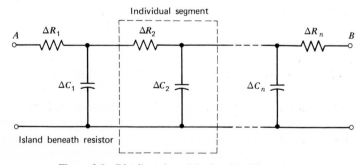

Figure 2.3 Distributed model of a thin-film resistor.

layer beneath the resistor. Note that for the segmented thin-film resistor model, ΔC is independent of the applied voltage. Furthermore, the capacitance per unit area of a thin-film resistor is less than that of a diffused resistor. A 0.5 mil wide $100 \, \Omega/\square$ diffused resistor typically has

$$\frac{\Delta C}{\Delta R} = 6.6 \times 10^{-2} \, \text{pF/k}\Omega \qquad (2.7)$$

A comparable thin-film resistor typically has

$$\frac{\Delta C}{\Delta R} = 4 \times 10^{-2} \, \text{pF/k}\Omega \qquad (2.8)$$

The typical temperature coefficients for commonly used thin-film resistors are much smaller than those for diffused resistors for temperatures between -55 and $+125°C$, as Figure 2.2 indicated. Typical values range from 50 ppm/°C for NiCr to approximately 600 ppm/°C for high-resistivity CrSi films. These TCs are an order of magnitude better than 10^3 ppm/°C, which is typical for diffused resistors.

The tolerance associated with the absolute value of both diffused and thin-film resistors is determined by the accuracy of reproducing the number of squares (the geometry) and the consistency and repeatability of the sheet resistance (diffusion or thin-film deposition). Geometric variations are due principally to variations in the resistor width. For a typical 0.5 mil wide resistor, a tolerance of $\pm 10\%$ is routinely obtained. To achieve the best possible matching between two or more resistors, widths of 1 or even 2 mils are sometimes employed. Resistivity variations are attributable to variations in the control of the impurity diffusion or thin-film deposition; a $\pm 15\%$ tolerance for the sheet resistance is typical. Resistivity and geometric variations must be considered to be independent variables when determining the total resistor tolerance. The typical percentages quoted represent variations achieved in at least 90% of all circuits in all locations on all wafers processed over long periods.

Much closer tolerance control is possible in matched resistors in monolithic circuits. Closeness of resistor matching is determined by the variation in geometry and resistivity between two very closely spaced (typically <20 mils) components in the same chip. Because of the physical proximity of matched IC components, gradual variations in physical processes that cause the relatively large variations in the resistor's absolute value over an entire collection of wafers, are much smaller for two matched resistors. For example, resistors 0.5 mil wide can easily be matched to better than 0.5% within the monolithic chip, whereas chip-to-chip variations may exceed 10%.

2.3 ANALOG INTEGRATED CIRCUIT CAPACITORS

Junction-Capacitor Models

The capacitance associated with a reverse-biased P–N junction is a distributed effect that is best modeled by a segmented network. A cross-sectional view of a junction capacitor is given in Figure 2.4 together with the segmented model. The N+ buried layer can be considered to be a conductor having negligible resistance. The substrate can also be considered to be an equipotential region because the bottom of the chip is contacted where it is attached to the IC package. Consequently, the

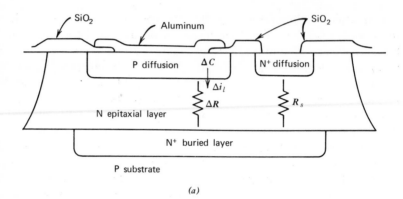

(a)

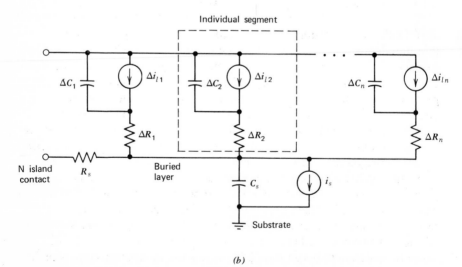

(b)

Figure 2.4 PN junction capacitor. (*a*) Cross-sectional view. (*b*) Distributed element model.

substrate parasitics of C_s and i_s are lumped elements because they connect two equipotential surfaces.

One segment of the distributed junction capacitance can be calculated from

$$\Delta C = \Delta A_j \left(\frac{q \varepsilon N}{2V} \right)^{1/2} \tag{2.9}$$

which is similar in form to Equation 1.8. The total effective capacitance is the sum of the individual ΔC segments. The ΔA_j term, which represents the area of one segment of the distributed model, can be found from Equation 2.2.

The additional elements in the junction-capacitor model account for various parasitic effects. The current source Δi_l is the junction-leakage current whose value at 25°C can be found from

$$\Delta i_l = 1 \text{ pA} \times \Delta A_j \tag{2.10}$$

where ΔA_j is in units of square mils. The ΔR component represents the segmented resistance from the P–N junction to the buried layer. The resistance from the buried layer to the N+ island contact is given by the lumped resistance R_s.

An important consideration of P–N junction capacitors at high-impedance circuit points, discussed in Chapter 3, is the leakage resistance. A typical junction capacitor at 125°C and 10 V bias has a leakage current of 150 nA. This leakage current is related to the applied voltage by

$$i_l = KV^{1/2} \tag{2.11}$$

where K is a proportional constant. The junction capacitance has an effective ac conductance g_l found by

$$g_l = \frac{di_l}{dV} = \tfrac{1}{2} KV^{-1/2} = \tfrac{1}{2} (KV^{1/2}) V^{-1} \tag{2.12}$$

$$g_l = \left(\frac{i_l}{2V} \right) = \frac{1}{z_l}$$

Using our typical parameters, we find

$$z_l = \frac{2 \times 10}{1.5 \times 10^{-7}} = 130 \text{ M}\Omega \tag{2.13}$$

This value of z_l seems to be exceptionally large, but it can be a significant impedance in high-performance circuitry.

MOS Capacitor Models

The oxide capacitor used in analog integrated circuits has insignificant leakage current between the capacitor "plates." Furthermore, the value of an MOS capacitor does not depend on a biasing voltage, and it has a much smaller parasitic series resistance than the P–N junction capacitor. A useful model for representing an MOS capacitor is presented in Figure 2.5. The distributed capacitance ΔC is given by

$$\Delta C = \frac{\Delta A_j \varepsilon}{t} \tag{2.14}$$

where ΔA_j represents a small segment of the total plate area. The ΔR resistance shown in the model represents the resistance of the portion of

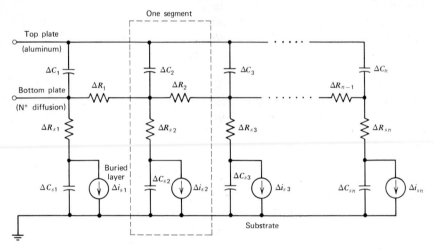

Figure 2.5 Distributed model for MOS capacitor.

the N+ bottom-plate diffusion beneath ΔA_j. Since the N+ diffusion produces a region of very low resistivity, ΔR is quite small and is usually considered to be a short circuit. The Δi_s quantity is the leakage current across the isolation region. The ΔR_s in the model represents the resistance from the N+ bottom-plate section, ΔA_j, to the buried layer. Last, ΔC_s is the buried-layer–substrate capacitance of the segmented area ΔA_j.

When the ΔR resistance is neglected, the distributed MOS capacitor model can be simplified to the lumped-element model of Figure 2.6. The capacitor C represents the total MOS capacitor, and R_s accounts for the total parasitic resistance from the bottom plate to the buried layer. (R_s is the parallel equivalent of the individual ΔR_s segments.) The capacitor C_s is the total parasitic capacitance between the isolation region and the substrate, and the current i_s is the total leakage current from the isolation region to the substrate.

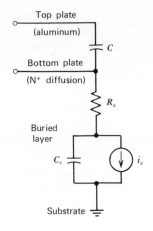

Figure 2.6 Simplified lumped-element model of MOS capacitor. The parasitic elements C_s and i_s are negligible in DI circuits.

In summary, a typical junction capacitor has a capacitance per unit area of

$$\frac{C}{A} = 2 \times 10^{-2} \text{ pF/mil}^2 \tag{2.15}$$

and a $V^{-1/2}$ bias voltage dependance. A typical MOS capacitor has

$$\frac{C}{A} = 0.13 \text{ pF/mil}^2 \tag{2.16}$$

and no voltage dependance. Furthermore, a 20 pF junction capacitor will have an associated leakage current of about 150 nA at 125°C, whereas an MOS capacitor of the same value will have negligible leakage current. Finally, the parasitic resistance of the MOS capacitor is much smaller than that of the junction capacitor. For these reasons, the MOS capacitor is favored wherever circuit performance is critical.

2.4   TRANSISTORS MODELS

Bipolar Transistors

The familar diode equation provides the most useful relationship for modeling bipolar transistors. This equation relates the emitter current I_E to the base–emitter voltage V_{BE} as

$$I_E = J_s A_E e^{\Lambda V_{BE}} \tag{2.17}$$

where

$$\Lambda = \frac{q}{kT} \tag{2.18}$$

The quantity J_s is the emitter saturation current density; A_E is the emitter area; k is the Boltzmann constant; and T is absolute temperature. For comparative purposes, at room temperature of 25°C or 298°K, we have

$$\frac{1}{\Lambda} = \frac{kT}{q} \approx 26 \text{ mV} \tag{2.19}$$

The emitter saturation current density J_s is a characteristic of the processing used to fabricate the transistor and is independent of the topology of the device. All devices made with the same diffusion (e.g., all NPNs in a monolithic circuit) will have very closely matched values of J_s. This extremely useful property for matching transistors forms the basis for many special circuit configurations peculiar to analog integrated circuits, as Chapter 3 reveals.

Solving Equation 2.17 for V_{BE} gives

$$V_{BE} = \frac{kT}{q} \ln \frac{I_E}{J_s A_E} \tag{2.20}$$

The current density J_s has a strong dependence on temperature, doubling for each 10°C increase in temperature. The J_s dependence, together with the absolute temperature term T, causes

$$\frac{\partial V_{BE}}{\partial T} \approx -2 \text{ mV/°C} \tag{2.21}$$

A second basic relationship for modeling a transistor is the dc, common-emitter, current gain h_{FE} (or β_{DC}), defined as

$$h_{FE} = \frac{I_C}{I_B} \tag{2.22}$$

where I_C and I_B are collector and base currents, respectively. The common-base current gain, α, is the ratio of collector to emitter current, or

$$\alpha_{DC} = \frac{I_C}{I_E} \tag{2.23}$$

These dc relationships are used primarily for calculating quiescent currents and voltages. We now illustrate small-signal, ac models useful for predicting a circuit's response to an ac signal. The most useful small-signal transistor model for low-frequency design and analysis is the common-base, h-parameter model.* The h-parameter model is particularly useful because its parameters can be measured with reasonable ease

---

* Formulas for relating the four h-parameters to the common-base, common-emitter, and common-collector configurations can be found in Millman and Halkias (12).

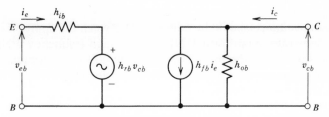

Figure 2.7 Small-signal, low-frequency h-parameter model for a transistor in the common-base (CB) configuration.

and can be correlated quite well to device structure. Furthermore, the model gives results consistent with experimental data. In the basic model (Figure 2.7), the hybrid parameters are defined as

$$h_{ib} \triangleq \left. \frac{v_{eb}}{i_e} \right|_{v_{cb}=0} \approx \frac{kT}{qI_E} \quad (\Omega) \tag{2.24}$$

$$h_{rb} \triangleq \left. \frac{v_{eb}}{v_{cb}} \right|_{i_e=0} \quad \text{(dimensionless)} \tag{2.25}$$

$$h_{fb} \triangleq \left. \frac{i_c}{i_e} \right|_{v_{cb}=0} = \frac{-\beta}{\beta+1} \quad \text{(dimensionless)} \tag{2.26}$$

$$h_{ob} \triangleq \left. \frac{i_c}{v_{cb}} \right|_{i_e=0} \quad \text{(mhos)} \tag{2.27}$$

Note that h_{ib}, an ac small-signal parameter, can be related to the dc emitter current I_E, as shown in Equation 2.24. For moderate currents, h_{rb} is independent of emitter current. However, both h_{rb} and h_{ob} vary with the dc biasing voltage between the base and collector according to $(V_{cb})^{-1/2}$. The h_{ob} parameter is directly proportional to the transistor collector current.

High-frequency effects in monolithic transistors can be considered by adding a capacitor C_{eb} to represent the emitter–base junction capacitance and a second capacitor, C_{bc}, to represent the collector–base junction capacitance to the hybrid model as in Figure 2.8. As a further refinement

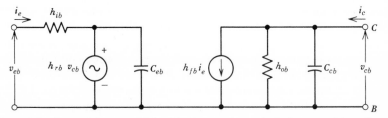

Figure 2.8 Common-base model for bipolar transistors, including capacitive effects.

of this high-frequency model, a capacitor to represent the island-to-substrate capacitance should be connected between the appropriate node and the substrate. For normal NPNs and high-frequency PNPs, the capacitor is added to the model between the collector terminal and the substrate (ac ground). For lateral PNPs, the capacitor is added between the base terminal and the substrate.

Bipolar Transistor Matching

Two or more integrated components on a monolithic chip will be matched to closer tolerances than randomly selected discrete components. This close matching is due to the proximity of integrated devices. Good thermal conductivity among monolithic devices is also a characteristic of integrated circuit components. Component matching in ICs forms the basis for many design techniques.

The most important characteristic to be controlled between two or more transistors is the V_{BE} match. Equation 2.20 shows that differences in J_s and A_E will contribute to component variation. The current density J_s depends on the diffusion process used to form the device; A_E depends on the accuracy of the formation of the aperture for the emitter diffusion. For typical device geometries, differences in emitter areas (ΔA_E) contribute more to component mismatch of V_{BE} in two or more monolithic devices than the mismatch due to ΔJ_s. A typical ΔV_{BE} mismatch for a normal NPN transistor pair is 0.5 mV at a current level of 0.01 to 1.0 mA. The second important transistor parameter to be matched among devices is β. Good β matching is achieved by matching the consistencies of the material in which the device is built and the consistency of the diffusions used to form the device. Typical IC matching of β is approximately 15% over a wide temperature range between −55 and +125°C.

Field-Effect Transistor Models

Both JFETs and MOS FETs are used in analog ICs. The basic current–voltage equations describing both types of devices are quite similar. However, some important differences must be considered in many applications.

JFETs always operate as depletion-mode devices. In this mode, drain current exists with zero gate-to-source voltage, V_{GS}. The drain current will become insignificantly small at some gate-to-source voltage called the pinchoff voltage V_P. At this voltage, the depletion layer has spread outward from the gate, completely filling the channel. When this layer completely fills the channel, current flow ceases because the channel is completely blocked.

MOS devices in analog ICs can operate in either the enhancement mode or the depletion mode. In the enhancement mode, drain current is zero for a zero gate-to-source voltage. The enhancement-mode device will begin to conduct drain current when the gate-to-source voltage reaches a threshold voltage, V_T. Further increases in the magnitude of V_{GS} cause corresponding increases in the device's drain current. Enhancement-mode N-channel devices have a positive V_T, but V_T is negative for P-channel devices.

The common processes used to make P-channel devices in ICs yield only enhancement-mode devices. N-channel MOS devices can be made either enhancement-mode or depletion-mode depending on the concentration of the body doping. Increasing this doping level increases the threshold voltage (enhancement mode), whereas decreasing the doping lowers V_T and will eventually cause depletion-mode operation.

The gate-to-source voltage of an MOS device can assume either polarity without damaging the device or causing circuit malfunction. Junction FETs however must always maintain a gate voltage that keeps the gate-to-source junction reverse-biased. Reverse biasing prevents the flow of excessive gate current and maintains a very high input impedance level. Keeping the junction reverse-biased requires a positive gate-to-source voltage for a P-channel device and a negative V_{GS} for an N-channel device.

The basic current-voltage relationship used to describe an MOS FET of either N- or P-type which operates in the saturated region ($V_{DS} > V_T$) is

$$I_D = \frac{wK'}{l}(V_{GS} - V_T)^2 \tag{2.28}$$

where w is the channel width, l is the channel length, and K' is a conductance parameter whose value depends on the fabrication process used to produce the FET and which is independent of device topology. For a depletion-mode device, the threshold voltage V_T in Equation 2.28 should be replaced by the pinchoff voltage V_P.

The most useful parameter for describing the small-signal operation of either an MOS or JFET is the transconductance g_m, given by

$$g_m = \frac{\partial I_D}{\partial V_{GS}}\bigg|_{V_d=0} = \frac{2wK'}{l}(V_{GS} - V_T) \tag{2.29}$$

A small-signal model for the FET using the g_m parameter appears in Figure 2.9. This equivalent circuit model is for an FET device operating with common source and body, which is the usual FET configuration employed in analog applications.

For a JEFT, C_{gs} represents the P–N junction capacitance between the

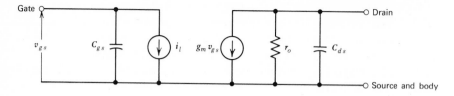

Figure 2.9 Small-signal model of FET.

gate and the source side of the channel. The C_{ds} capacitance accounts for the gate-junction capacitance on the drain side of the channel, and i_l is the junction-leakage current.

For a MOS FET, C_{gs} represents effective parasitic capacitance between the gate metal and the source and body regions of the MOS device. The C_{ds} capacitance accounts for the P–N junction capacitance of the drain-to-body P–N junction.

The resistance r_o is the small-signal, common-source output impedance of the FET. The most important parameter that serves as an index of matching between FET devices is V_{off}. This is the differential voltage between the gates of two matched FET devices carrying equal drain currents. Ideally, V_{off} would be zero. However, some finite offset voltage does exist because of inaccuracies of gate topology control and lack of uniformity of the diffusion and oxidations used to form the devices. Typical FET parameters are $V_{off} \approx 10$ mV for current levels between 0.01 and 1.0 mA.

A basic conflict exists when one tries to optimize the device topology of an FET. The gate capacitance is proportional to the gate area ($l \times w$), whereas the transconductance is proportional to the channel width-to-length ratio (w/l). To minimize the capacitance per unit g_m requires minimizing

$$\frac{C}{g_m} = \frac{K_1(l \times w)}{K_2(w/l)} = K_3 l^2 \qquad (2.30)$$

where the K's are conversion constants. Thus Equation 2.30 is minimized by keeping the channel lengths as short as possible. On the other hand, optimum matching between FET devices requires consistent channel geometries. This is achieved by utilizing the largest possible value for the smallest device dimension, which unfortunately is the channel length. Thus a device design dilemma is precipitated by the need for small channel lengths to minimize the C/g_m ratio and the large channel lengths required for better device matching.

2.5 INTEGRATED DIODES

Integrated diodes can be formed using any of the P–N junctions available in a monolithic circuit structure. However, only two junctions in the basic NPN structure are readily suitable for circuit applications. These junctions are the base–emitter and base–collector junctions. The collector–substrate diode is not useful because its anode, the substrate, is connected to the most negative voltage in the circuit. Figure 2.10 illustrates five possible diode connections using either/or both the base–emitter and base–collector junctions.

The series, bulk resistances between the device terminals and the actual diode junctions represent the most significant parasitics. With respect to

Diode Connection	Identification	Series Resistance	Reverse Breakdown	Parasitic[a] PNP
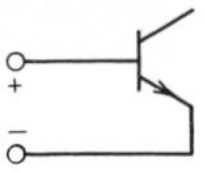	$I_c = 0$	Low $(\approx r_b)$	Low $(\approx 7\,\text{V})$	No
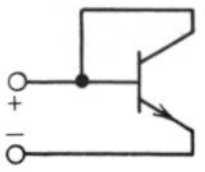	$V_{BC} = 0$	Low $(\approx r_{cs} + r_b/\beta_0)$	Low $(\approx 7\,\text{V})$	No
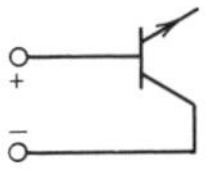	$I_E = 0$	High $(\approx r_b + r_{cs})$	High $(>40\,\text{V})$	Yes
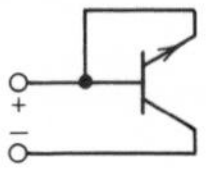	$V_{BE} = 0$	High $(\approx r_b + r_{cs})$	High $(>40\,\text{V})$	Yes
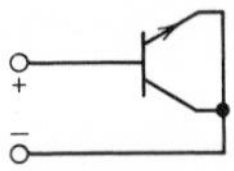	$V_{CE} = 0$	High $(\approx r_b + r_{cs})$	High $(>7\,\text{V})$	Yes

[a] These parasitic transistors occur only in JI circuits. No parasitic bipolar transistors are possible in DI.

Figure 2.10 Comparison of practical diode connections for an NPN transistor.

bulk resistance, the $V_{BC} = 0$ diode has the lowest possible value because any parasitic resistance in the base terminal appears divided by the β of the transistor. For this reason, this connection is the most widely used diode configuration in integrated circuits.

2.6 TYPICAL DEVICE PARAMETERS

In Figure 2.11 we have summarized typical device parameters for both bipolar and FET transistors. For comparative purposes the parameters of the various bipolar devices should be referenced to the normal NPN transistor. As the data indicate, a special improvement in one device parameter is usually accompanied by a degradation of one or more of the other parameters.

		Bipolar Devices					
Parameter	Q-Point	Normal NPN	Super-β NPN	Lateral PNP	Substrate PNP	High-Frequency NPN	High-Frequency PNP
β^a	$V_{CE} = 10\text{ V}$ $I_C = 0.1\text{ mA}$	350	2500	15	125	350	150
BV_{CEO}	$\pm 10\,\mu\text{A}$	55 V	7 V	60 V	60 V	55 V	60 V
BV_{CBO}	$\pm 10\,\mu\text{A}$	100 V	15 V	100 V	100 V	80V	75 V
BV_{EBO}	$\pm 10\,\mu\text{A}$	6.3 V	6.6 V	100 V	100 V	6.5 V	7.0 V
f_T^b	$V_{CE} = 10\text{ V}$ $I_C = 2\text{ mA}$	450 MHz	550 MHz	1 MHz	15 MHz	850 MHz	650 MHz
h_{ob}	$V_{CE} = 10\text{ V}$ $I_C = 0.1\text{ mA}$	0.005 μmho	0.009 μmho	0.1 μmho	0.03 μmho	0.005 μmho	0.012 μmho
h_{rb}^c	$V_{CE} = 10\text{ V}$ $I_C = 0.1\text{ mA}$	3×10^{-4}	3×10^{-3}	5×10^{-3}	5×10^{-3}	3×10^{-4}	2×10^{-4}

^{a} $h_{fb} \triangleq -\beta/(\beta + 1)$.
^{b} f_T varies with collector current I_C. The values quoted are peak values.
^{c} The fourth h-parameter, h_{ib}, is given by Equation 2.24.

		Field-Effective Devices		
Parameters	Q-Point	P-channel JFET	P-channel MOS FET	N-channel MOS FET (depletion mode)
g_m	$I_D = \pm 2\text{ mA}$ $V_{SD} = \pm 5\text{V}$	500 μmho	500 μmho	1000 μmho
BV_{d-sb}	$\pm 10\,\mu\text{A}$	6.5 V	50 V	40 V
V_T	$I_D = \pm 1\,\mu\text{A}$ $V_{SD} = \pm 5\text{ V}$	+2.5 V	−3.5 V	−1 V

Figure 2.11. Typical device parameters.

Some general observations not reflected in Figure 2.11 are appropriate here:

1. Lateral and substrate PNP devices have excellent β temperature coefficients. Typical changes for β are approximately 10% per 100°C temperature change. β changes for other transistors in the table is typically 90% per 100°C.

2. The offset voltage between two matched super-β devices is poorer than for the other transistors. Typical values are ~ 5 mV for the super-β and 0.5 mV for the remaining devices.

3. High-frequency transistors require the DI process and extra diffusions as discussed in Chapter 1.

4. Lateral and substrate PNPs have poor h_{ob} and h_{rb} parameters which limit their usefulness in high-impedance circuitry.

2.7 COMPUTER UTILIZATION FOR CIRCUIT ANALYSIS AND DESIGN

Analysis and design of analog ICs soon become difficult when a number of individual components are interconnected and modeled using the previously presented equivalent circuits. Furthermore, as is generally the case when new circuit designs are being developed, the models of the individual devices are pushed to new limits where their accuracy becomes questionable. Thus more sophisticated models are needed, which in turn make manual calculation techniques very difficult and of dubious accuracy. Utilization of the computer as a design and analysis tool has sparked much recent interest, particularly during the last decade. The computer represents a powerful design tool for simulation of circuit response. After the inevitable initial programming problems have been resolved and after correlation has been achieved with the computer simulation and an experimental circuit bread board, the design engineer can proceed to predict and optimize circuit behavior with a high degree of confidence, using the computer. Computer simulation is a particularly efficient way for determination of circuit sensitivities from variations of specific network parameters. However, important practical programming considerations must be resolved before the computer can be an effective analysis tool.

The primary consideration to be made in selecting a program is the required compromises among circuit size, model complexity, and the type and accuracy of the information required. For example, a specific analysis program such as ECAP* produces adequate results (accuracy $\approx 5\%$) for

*ECAP: *Electronic Circuit Analysis Program,* developed by IBM in 1965.

dc and small-signal ac characterization for circuits having a relatively large number of components (typically as many as 50 active devices and 100 passive devices). However, fairly simple device models must be used with ECAP, limiting the accuracy of this program for high-frequency and large-signal applications. Because of this limitation, the circuit design engineer may prefer to use manual calculations for the complete circuit wherever accuracy is not critical.

Other computer simulation programs, such as SCEPTRE,* utilize a more generalized active device model that makes the program useful for both large- and small-signal transient analysis. Furthermore, SCEPTRE will perform dc calculations. A major limitation of SCEPTRE is the number of circuit nodes the program can handle on a given computer. For example, a SCEPTRE program run on the UCC 1108 system from a COPE† 45 time-shared terminal is limited to a circuit with between 20 and 25 transistors. Since many analog ICs contain more than 75 active devices, this nodal limitation represents a serious constraint.

More efficient computer programs have been developed such as AEDCAP‡ and SPICE,§ which can handle a large number of circuit components. However, only a limited number of programs are well suited for analog IC design.

A very important consideration of a design engineer is the relative difficulty or ease with which he can interface with the computer. For example, if an engineer must repeatedly submit a card deck to a service terminal and then wait overnight, only to discover that the program did not run because of a simple format error, he may abandon the computer and resort to long and tedious hand calculations for results. However, if the engineer can interface with the computer on a real-time basis, his usage of the computer as a design tool will be considerably enhanced.

2.8   CONCLUSIONS

In this chapter we have presented a number of equivalent circuit models for the integrated devices fabricated in the first chapter. Constant

---

* SCEPTRE: System for Circuit Evaluation and Prediction of Transient Radiation Effects, developed by IBM in 1967.

† COPE: Computer Operated Peripheral Equipment, developed by UNIVAC in 1970.

‡ AEDCAP: Automated Engineering Design Circuit Analysis Program, developed by Softech in 1971.

§ SPICE: Simulation Program with Integrated Circuit Emphasis, developed at the University of California, Berkeley in 1970.

attempts were made while developing these models to relate the physical properties of the actual device to specific portions of the device model. Models were given for diffused and thin-film resistors, junction and MOS capacitors, and bipolar and field-effect transistors. All models were considered from the viewpoint of their application as elements in analog integrated circuits.

BIBLIOGRAPHY

1. Ahmed, H., and P. J. Spreadbury, *Electronics for Engineers, An Introduction,* Cambridge University Press, London, 1973.
2. Angelo, E. James, Jr., *Electronics: BTJs, FETs, and Microcircuits,* McGraw-Hill, 1969.
3. Belove, Charles, Harry Schacter, and Donald L. Schilling, *Digital and Analog Systems, Circuits, and Devices: An Introduction,* McGraw-Hill, New York, 1973.
4. Chirlian, Paul M., *Integrated and Active Network Analysis and Synthesis,* Prentice-Hall, Englewood Cliffs, N.J., 1967.
5. Deboo, Gordon J., and Clifford N. Burros, *Integrated Circuits and Semiconductor Devices: Theory and Application,* McGraw-Hill, New York, 1971.
6. Fitchen, Franklin C., *Electronic Integrated Circuits and Systems,* Van Nostrand-Reinhold, New York, 1970.
7. Gray, Paul E., and Campell L. Searle, *Electronic Principles: Physics, Models, and Circuits,* Wiley, New York, 1969.
8. Grebene, Alan B., *Analog Integrated Circuit Design,* Van Nostrand-Reinhold, New York, 1972.
9. Huelsman, L. P., *Active Filters: Lumped, Distributed, Integrated, Digital and Parametrics,* McGraw-Hill, New York, 1970.
10. Hunter, Lloyd P., *Handbook of Semiconductor Electronics,* McGraw-Hill, New York, 1970.
11. Meyer, Charles S., David K. Lynn, and Douglas J. Hamilton, *Analysis and Design of Integrated Circuits,* McGraw-Hill, New York, 1968.
12. Millman, Jacob, and Christos C. Halkias, *Electronics Devices and Circuits,* McGraw-Hill, New York, 1967.
13. Millman, J., and C. C. Halkias, *Integrated Electronics: Analog and Digital Circuits and Systems,* McGraw-Hill, New York, 1972.
14. Motchenbaucher, C. D., and F. C. Fitchen, *Low Noise Electronic Design,* Wiley, New York, 1973.
15. Pierce, J. F., and T. J. Paulus, *Applied Electronics,* Merrill, Columbus, Ohio, 1972.
16. Pierce, J. F., *Semiconductor Junction Devices,* Merrill, Columbus, Ohio, 1967.
17. RCA Inc., "Linear Integrated Circuits," RCA Corp., Technical Series IC-42, 1970.
18. RCA Inc., "Linear Integrated Circuits and MOS Devices," RCA Corp. No. SSD-202, 1972.
19. Ryder, John D., *Electronic Fundamentals and Applications,* Prentice-Hall, Englewood Cliffs, N.J., 1970.

20. Schilling, Donald L., and Charles Belove, *Electronic Circuits: Discrete and Integrated*, McGraw-Hill, New York, 1968.

21. Schwartz, Seymour, *Integrated Circuit Technology*, McGraw-Hill, New York, 1967.

22. Warner, Raymond M., Jr., and James N. Fordemwalt, *Integrated Circuits.*, McGraw-Hill, New York, 1965.

CHAPTER 3

ANALOG INTEGRATED CIRCUIT CONFIGURATIONS

An ever-increasing number of analog integrated circuits are available for accomplishing a multitude of signal-processing operations. All analog IC devices can be categorized according to design and application into one of two broad classifications. The first grouping includes analog IC devices designed for very specific applications. By far the largest subclass of this group are analog IC devices intended for communications systems such as video amplifiers, chroma demodulators, and stereo multiplexers. Some other examples of specialized, noncommunications-type devices in the first grouping are voltage regulators, line drivers and receivers, and sense amplifiers.

The second general category is comprised of all analog IC devices whose designs are intended to serve a large variety of applications. The largest subclass of this grouping is the operational amplifier; also included are voltage comparators, analog multipliers, phase-locked loops, and integrated power amplifiers.

All analog ICs, regardless of their design classification, are composed of recognizable circuit structures that accomplish specific functions within the device. The building-block structures are not revolutionary new circuits but instead are well-known configurations whose designs are based on the fabrication advantages and limitations generated by the IC manufacturing process.

In this chapter our first intention is to describe these basic circuit configurations, citing important philosophies reflected in their designs. The specific circuits to be examined are differential amplifiers, constant-current sources, biasing networks, current mirrors, emitter followers, gain stages, and output drivers. We will indicate practical limitations inherent in these basic circuit structures and show how many of the shortcomings can be circumvented by minor modifications to the basic circuits.

Our second objective is to describe several more specialized circuit structures that have unique features for removing some inherent fabrication or design limitation that cannot be eliminated by minor modifications. Also in this chapter we consider circuits designed to achieve specialized subsystem functions such as analog multiplication and analog switching.

BASIC BIPOLAR STRUCTURES

3.1 DIFFERENTIAL AMPLIFIERS

The differential amplifier circuit is the single most important building block in analog IC designs. Since a differential amplifier produces an amplified output signal linearly related to the difference between two input signals, the circuit is frequently called a difference amplifier.

Figure 3.1 is a symbolic representation of a differential amplifier having two input-signal voltages, v_1 and v_2, and one output voltage, v_O. The output of an *ideal* differential amplifier is linearly related to the input difference voltage v_D, through the difference-mode gain A_d, or

$$v_O = A_d(v_2 - v_1) = A_d v_D \tag{3.1}$$

Thus we see that any signal common to both inputs will be removed because of the subtraction and will have no effect on the output voltage. However, this ideal operation cannot be achieved in practice because the output voltage will also depend to some extent on the common-mode

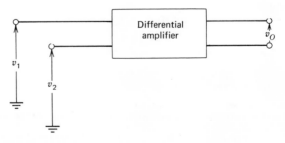

Figure 3.1 Block diagram of an ideal differential amplifier whose output is a linear function of the differential voltage signal: $V_O = A_d(V_2 - V_1)$.

characteristics of the amplifier and the input signal voltage according to

$$v_O = A_d v_D + A_c v_C \qquad (3.2)$$

where A_c is the common-mode voltage gain and v_C is the average voltage level of two input signals, or

$$v_C = \tfrac{1}{2}(v_2 + v_1) \qquad (3.3)$$

A figure of merit used to evaluate the performance of a practical differential amplifier is the common-mode rejection ratio, defined as

$$\text{CMRR} = \rho = \left| \frac{A_d}{A_c} \right| \qquad (3.4)$$

Obviously, the larger the CMRR, the closer the practical differential amplifier approaches the ideal case where $A_c = 0$.

The basic circuit structure for a differential amplifier is depicted in Figure 3.2. (The I_0 source represents a constant current whose value is independent of all external circuit influences. We describe the design of this circuit and its implementation in Section 3.3.) The unique properties achievable with differential amplifiers are due mainly to the symmetry between the two sides of the circuit. Being balanced, this configuration is ideally suited for integrated circuit fabrication because of the close device-matching and tracking characteristics inherent in monolithic components.

In beginning our analysis of the basic differential amplifier stage of Figure 3.2, we will assume that both resistors and both transistors are precisely matched, with the I_0 current splitting equally between the two circuit halves. Also, if we assume equal input voltages ($v_1 = v_2$), we have a truly symmetrical network having two sides that are mirror images. This type of network is best analyzed using the bisection theorem whose proof may be found in the literature (10, 13). Because of the circuit symmetry, this theorem allows the analysis of only one-half of the total circuit (see Figure 3.3). Using conventional circuit analysis, the common-mode voltage gain can be found as

$$A_c = \frac{v_O}{v_C} = -\frac{g_m R_c}{1 + 2g_m R_0} \qquad (3.5)$$

where $2R_0$ is the impedance associated with the current source ($I_0/2$) and g_m is the transconductance of Q. (Thus producing an ideal difference amplifier first requires the creation of an ideal current source.) The transistor's transconductance g_m can be found from

$$g_m = \frac{I_0}{2} \Lambda \qquad (3.6)$$

where Λ is the quantity defined in Equation 2.19.

The basic differential amplifier of Figure 3.2 becomes an asymmetric

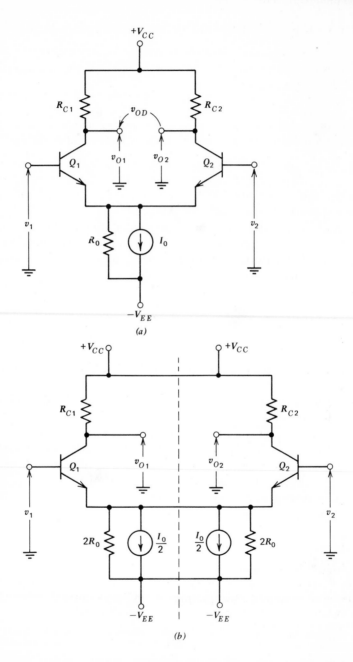

Figure 3.2 Circuit diagram of a basic differential amplifier stage. (*a*) Before bisection. (*b*) After bisection.

56

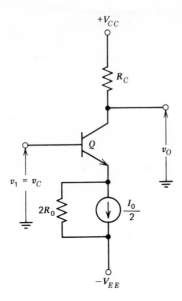

Figure 3.3 Equivalent half-circuit for common-mode input signal.

network when the two input voltages are equal in amplitude and 180° out of phase, or when $v_1 = -v_2$. For this input condition, the voltage level at the common-emitter point remains constant because positive changes in the base voltage of one transistor will be exactly counterbalanced by negative changes in the other transistor's base voltage. Thus the common-emitter point appears as a virtual ground, and the simplified circuit of Figure 3.4 can be used to determine the difference-mode gain for the circuit. Using conventional circuit analysis, the difference-mode gain can be expressed as

$$A_d = \frac{v_O}{v_D} = \frac{g_m R_C}{2} \qquad \text{(single-ended output)} \qquad (3.7)$$

If the output voltage is not taken referenced to ground but taken instead between the collectors of the two transistor as indicated by v_{OD} in Figure 3.2, this differential gain value is twice as large as the expression given in Equation 3.7, or

$$A'_d = \frac{v_{OD}}{v_D} = g_m R_C \qquad \text{(differential output)} \qquad (3.8)$$

The CMRR for single-ended operation reduces to

$$\rho = \frac{1 + 2g_m R_0}{2} \approx g_m R_0 \qquad (3.9)$$

and is normally expressed in decibels.

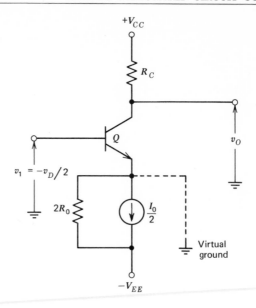

Figure 3.4   Equivalent half-circuit for different-mode input signal.

Many differential input stages are designed to be transconductance amplifiers. For this basic type of amplifier (Figure 3.5), the differential input voltage v_D generates a differential output current i_D of the form

$$i_D = i_1 - i_2 \tag{3.10}$$

Since Q_1 and Q_2 operate in their linear regions, we can write the following equations for their collector currents:

$$i_1 = I_S e^{\Lambda(v_D + v_{BE2})} \tag{3.11}$$

$$i_2 = I_S e^{\Lambda v_{BE2}} \tag{3.12}$$

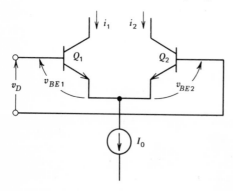

Figure 3.5   Basic transconductance differential amplifier.

where I_S is the reverse-saturation current. Let us now assume equal collector and emitter currents and recognize that

$$I_0 = i_1 + i_2 \tag{3.13}$$

After some algebraic manipulation, we can write

$$i_1 = \frac{I_0 e^{\Lambda v_D}}{1 + e^{\Lambda v_D}} \tag{3.14}$$

$$i_2 = \frac{I_0}{1 + e^{\Lambda v_D}} \tag{3.15}$$

Combining Equation 3.10 with the two previous equations yields

$$i_D = I_0 \frac{e^{\Lambda v_D} - 1}{e^{\Lambda v_D} + 1} = I_0 \tanh \frac{\Lambda v_D}{2} \tag{3.16}$$

This equation, graphed in Figure 3.6, shows that i_D increases monotonically with increasing input voltage v_D. Furthermore, the maximum differential current that can be generated by this basic input circuit is I_0. The I_0 current is the principal mechanism controlling the speed/power figure of merit of an amplifier. Increasing I_0 will increase the amplifier's gain (Equations 3.6 and 3.7), common-mode rejection ratio, and slew rate. However, increasing I_0 will have the undesirable effect of increasing the device's power dissipation. For these and other considerations, several common modifications are frequently added to the structure of the

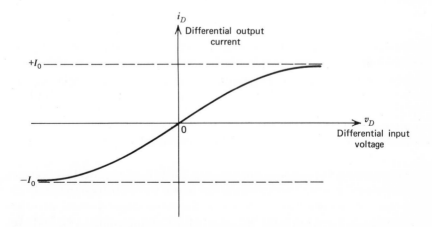

Figure 3.6   Conventional input-circuit transfer function.

differential amplifier to improve overall analog IC performance. We now illustrate a few popular modifications.

3.2 MODIFIED DIFFERENTIAL AMPLIFIER CIRCUIT STRUCTURES

By far the most critical block in an analog IC system is the input stage, whose performance has the dominant effect on the performance of the entire monolithic device. One important practical consideration of the input stage is the loading on the attached signal source(s) that is produced by this circuit. The extent of the loading effect is best reflected by the input-bias current and the input impedance. (The input-bias current is the average of the two dc currents that bias the two input transistors in a differential amplifier stage.) In some applications, the input-bias current and the input impedance may be too high and too low, respectively, to provide acceptable performance for the basic differential amplifier configuration of Figure 3.2a.

The most direct method for reducing the bias current and increasing the input impedance is to reduce the collector currents in the differential transistors. However, this approach has two serious detrimental effects. Since the g_m of the input transistors is directly proportional to the quiescent collector current (Equation 3.6), reducing I_C will decrease the stage's voltage gain. Furthermore, reduced current levels require larger-valued resistors, which in turn require more chip area. For these reasons, a Darlington configuration of the basic type appearing in Figure 3.7, is often used to reduce the bias current and increase the input impedance of the input stage. The additional Q_3 and Q_4 transistors buffer the normal differential Q_1 and Q_2 transistors from the v_1 and v_2 input signals. This buffering reduces the input-bias current of the basic differential stage by approximately a factor of $\beta_3/1 + 2\beta_1 I_{01}/I_0$. Furthermore, the input impedance of the Darlington configuration is increased by approximately this same factor. With the Darlington connection, typical bias currents of 5 to 10 nA and typical input impedances of 10 to 20 MΩ are achieved routinely.

One serious limitation of the Darlington configuration is the typical match between the composite Q_1–Q_3 and Q_2–Q_4 pairs is not as close as is usually achieved between single devices. Therefore, input-offset voltages and voltage drifts with temperature are worse for a Darlington configuration than for a basic, two-transistor differential stage. Typical performance indexes are ±3 mV for the input-offset voltage and ±15 μV/°C for the temperature coefficient of the offset voltage.

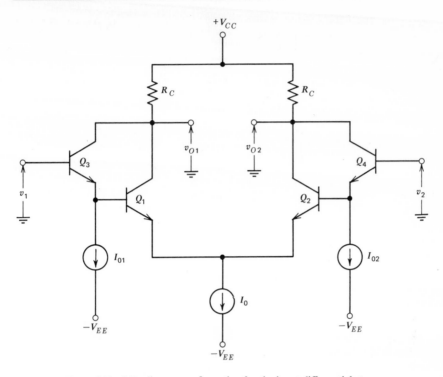

Figure 3.7 A Darlington configuration for the input differential stage.

Another limitation shared by the basic differential amplifier and the Darlington configuration is the ability of these circuits to respond to a large, differential input signal. A large signal will force the transistors on one side of the circuit to saturate while the corresponding transistors on the other side are cut off. The maximum output current available from the differential stage is $\pm I_0$ as in Figure 3.6. Since this output current usually charges a compensation capacitor, the maximum, large-signal rate of change of the voltage is limited to a constant value and is no longer proportional to v_D. This rate of change is defined as the slew rate.

Slew-rate performance in the differential input stage can be improved by placing two matched, small-valued resistors in the emitter circuits of both input transistors, as in Figure 3.8. This design modification, known as emitter degeneration, is a form of local, negative feedback that tends to lessen the nonlinearity of the input-diode characteristic of the V_{BE} junction. The addition of a linear resistance in series with a nonlinear diode tends to dilute the nonlinearity and to allow larger input signals to be applied with reduced output distortion.

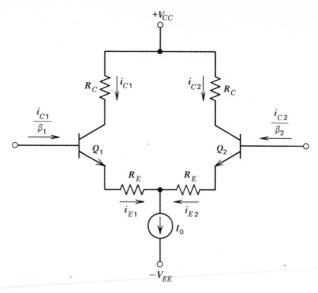

Figure 3.8 Modification of the basic differential amplifier stage through emitter degeneration.

The addition of the two emitter resistors R_E reduces the stage gain by reducing the transconductance in accordance with

$$g'_m = \frac{1}{1/g_m + R_E} \tag{3.17}$$

Solomon (27) has shown that slew rate is increased by the same factor by which the transconductance is reduced for a constant gain–bandwidth product since the value of the compensation capacitor can be reduced by a factor of g_m/g'_m.

Adding the two emitter resistors further degrades the input characteristics of the differential amplifier stage. Mismatches between corresponding transistors in the configuration become more apparent because additional error terms appear (e.g., $\Delta R_E I_E$ and $\Delta I_E R_E$, where ΔR_E and ΔI_E represent resistance and emitter-current mismatches).

A further refinement of the basic differential amplifier appears in Figure 3.9. This circuit incorporates an NPN–PNP type of Darlington input connection; emitter degeneration is provided by the two diodes. Addition of the diodes in the emitters of these input transistors increases the differential input capability of the stage by the reverse-breakdown voltage of the diodes. The NPN–PNP Darlington configuration increases the common-mode range of the stage for given supply voltages and provides

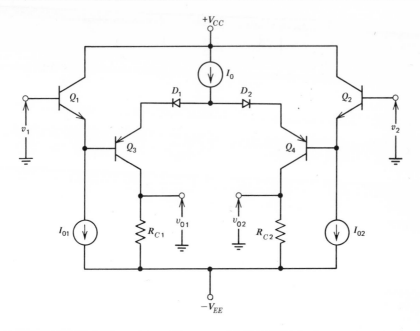

Figure 3.9 Modified differential amplifier stage with NPN–PNP Darlington connection and diode degeneration.

level shifting to the negative supply, while maintaining the input-bias current characteristics of the relatively high-β NPN transistors.

3.3 CONSTANT-CURRENT CIRCUITS

We have seen that a source of constant direct current is necessary for proper operation of the differential amplifier stages previously considered. In addition to their employment here, constant-current circuits are used as basic structures in numerous other analog IC configurations.

There are two basic types of constant-current circuits; circuits that *supply* a constant current to a load (current sources) and circuits that *remove* a constant current from a load (current sinks). The constant-current source generally employs PNP or p-channel devices, whereas constant-current sinks usually use NPN or n-channel devices. Because of the inherent advantages of NPN devices both in device fabrication and device performance, constant-current sinks are more typical than current sources. For this reason, our discussion focuses on current sinks, with the

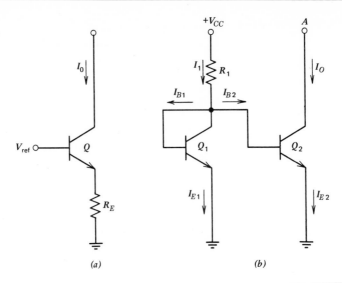

Figure 3.10 Constant-current circuits. (*a*) Basic transistor circuit. (*b*) Diode-biased circuit.

realization that our analysis can be applied to current sources simply by changing all NPN to PNP devices and reversing the polarity of all voltage supplies.

A single transistor and a resistor represent the most basic, constant-current configuration. (see Figure 3.10*a*). A constant dc reference voltage V_{ref} is maintained on the base of Q. The collector current is the current maintained constant and is given by

$$I_0 = \alpha \frac{V_{ref} - V_{BE}}{R_E} \qquad (3.18)$$

If $V_{ref} \gg V_{BE}$ and $\alpha \approx 1$, as is the usual case, the constant current can be approximated by

$$I_0 \approx \frac{V_{ref}}{R_E} \qquad (3.19)$$

The conductance associated with a constant-current circuit is the best indicator of circuit performance. An ideal constant-current circuit will have zero associated conductance (infinite shunt resistance). We can easily evaluate the output conductance of Figure 3.10*a* by considering that a modulating ac voltage ΔV is applied to the collector, causing I_0 to change by some small amount ΔI. Using the hybrid equivalent circuit of

Figure 2.7, the output conductance can be found as

$$g_{out} = \frac{\Delta V}{\Delta I} = h_{ob} - \frac{h_{fb}h_{rb}}{h_{ib} + R_E} \tag{3.20}$$

For comparative purposes, assume that $R_E = 500 \, \Omega$ and that the constant-current level desired is 0.1 mA. Using the typical device parameters for a normal NPN transistor given in Figure 2.11, g_{out} can be evaluated as

$$g_{out} = 5 \times 10^{-9} - \frac{(-0.997)(3 \times 10^{-4})}{259 + 500}$$

$$g_{out} = 5 \times 10^{-9} + 3.94 \times 10^{-7} \approx 4 \times 10^{-7} \text{ mho} \tag{3.21}$$

In this typical calculation, the conductance is equivalent to an output impedance of 2.5 MΩ, which is frequently too low for many critical differential amplifier and high-impedance applications. As the numerical calculation shows, the value of g_{out} is limited most by the h_{rb} term. In Section 3.11, we discuss neutralization, an advanced circuit design technique that reduces both the h_{rb} and h_{ob} terms.

A constant-current circuit designed to take advantage of analog IC characteristics appears in Figure 3.10b. Close device matching exists between transistor Q_1 (connected as a diode with $V_{BC}=0$) and Q_2. The current maintained constant is I_0. The base–emitter currents are proportional to the geometries of the transistors' emitter areas, or

$$\frac{I_{E1}}{I_{E2}} = \frac{A_{E1}}{A_{E2}} = \text{constant } K \tag{3.22}$$

Since the two emitter currents are matched through K, the circuit of Figure 3.10b is frequently referred to as a current mirror.

If the β's of both transistors are reasonably large (50 or greater), the base currents can be neglected with respect to the collector currents. The desired current I_0 is given by

$$I_0 = \frac{V_{CC} - V_{BE}}{R_1} \frac{A_{E2}}{A_{E1}} \approx \frac{V_{CC}A_{E2}}{R_1 A_{E1}} \tag{3.23}$$

Thus the current I_0 can be set through appropriate transistor geometry and by circuit design attention to V_{CC} and R_1. Usually the supply voltage V_{CC} is derived from a stable voltage source, covered in the next section. This circuit possesses some degree of thermal stability, since there is a small cancellation of thermal effects between R_1 and the base–emitter junction potentials due to the sign difference between the respective temperature coefficients.

A general expression for the output conductance at the collector of a transistor is derived in Appendix A, and the resulting equation is

$$g_{out} \approx h_{ob}\left(1 - \frac{h_{fb}R_B}{h_{ib} + R_E}\right) - \frac{h_{fb}h_{rb}}{h_{ib} + R_E} \tag{3.24}$$

Applying this equation to the basic current-mirror structure in Figure 3.10b gives

$$g_{out} \approx 2h_{ob2} + \frac{h_{rb2}}{h_{ib2}} \tag{3.25}$$

which assumes $R_B = h_{ib1} \ll R_1$, $h_{FE} \gg 1$, $R_E = 0$, and Q_1 and Q_2 are perfectly matched devices in typical designs. Again let us use the same typical device parameters cited previously to evaluate g_{out}. We have

$$
\begin{aligned}
g_{out} &\approx 2(5 \times 10^{-9}) + \frac{3 \times 10^{-4}}{259} \\
&\approx 1 \times 10^{-8} + 1.157 \times 10^{-6} = 1.167 \ \mu\text{mho}
\end{aligned}
\tag{3.26}
$$

As Equation 3.26 shows, the h_{rb2}/h_{ib2} term normally dominates the g_{out} expression and limits the typical output impedance achieved with the basic current mirror to approximately 1 MΩ.

One technique for lowering the output conductance is to add resistors in series with the emitters of both Q_1 and Q_2. These resistors increase the effective h_{ib} to the value

$$h'_{ib} = h_{ib} + R_E \tag{3.27}$$

If R_E is made very large, the h_{rb} term in Equation 3.25 can be reduced to a level approaching h_{ob}. However, for typical operating collector currents $\leqslant 300 \ \mu$A, R_E must be a large-valued resistor requiring excessive chip area and producing a large, intolerable voltage drop.

In a simple modification to the basic diode-biased, constant-current sink (Figure 3.11), an emitter resistor is added to the Q_2 transistor. Positioning a relatively small value of resistance here produces a circuit capable of sinking extremely low values of constant current, typically of a few microamperes. The voltage potential existing across the R_2 resistor is the difference between the base–emitter voltages of Q_1 and Q_2, or

$$R_2 I_{E2} = V_{BE1} - V_{BE2} \tag{3.28}$$

Since $I_{E2} \approx I_0$, the value of R_2 needed to set a current level of I_0 can be found from

$$R_2 = \frac{1}{\Lambda I_0} \ln \frac{I_1}{I_0} \tag{3.29}$$

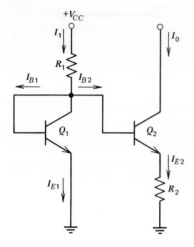

Figure 3.11 Diode-biased constant-current circuit with emitter resistor.

As a practical numerical example, suppose we desire a constant-current sink capability of 20 μA for a differential amplifier stage. To achieve this small current value with the circuit of Figure 3.10a would require an R_1 resistance value in the megohm range, which is impractical for integrated devices. (Practical considerations limit the area ratios possible through Equation 3.22 to the range between approximately 1 and 5.) The major limitation here is a loss of the matching and tracking characteristic between the two devices. However, by using the circuit of Figure 3.11 and setting $I_1 = 1$ mA, the required value of R_2 is a very practical size of 4.9 kΩ.

The stability of each of the basic constant-current circuits we have discussed is directly related to some biasing voltage, V_{ref}. This voltage is generally derived from an internal bias network that is stabilized against power-supply variations and temperature changes. We now illustrate how these different reference-voltage sources are designed and implemented through various bias networks.

3.4 BIAS NETWORKS

The dual of the constant-current circuit is the constant-voltage configuration, which is designed to present a stable voltage level in the presence of all disturbing influences. Two important influences in the design of voltage–source circuits are temperature and power-supply variations. A third and equally important consideration is the circuit's ability to maintain the required reference-voltage level under varying load–current conditions. To achieve good regulation, the voltage–source circuit must possess an extremely low output impedance. Fortunately in most analog ICs circuit insensitivity either to power-supply–temperature variations or to varying load conditions is the dominant concern. Usually some degree of design compromise must be reached, thereby minimizing the effect of only one of these influences.

One of the simplest circuits for maintaining a relatively fixed output voltage during load–current variations is presented in Figure 3.12. The low output impedance of the emitter-follower configuration is responsible for maintaining a reasonably constant output voltage under a variety of load conditions as represented by Z_L. The reference voltage V_{ref} is set in accordance with

$$V_{ref} \approx V_{CC} \frac{R_2}{R_1 + R_2} \tag{3.30}$$

The R_1–R_2 resistor string sets the reference-voltage level and provides first-order compensation against temperature variations. Additional temperature stability is supplied by the tracking between the diode voltage and the transistor's base–emitter junction potential. However, the major shortcoming of this circuit is its sensitivity to power-supply variations. Replacing resistor R_2 with a zener diode is helpful in removing this

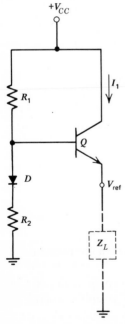

Figure 3.12 Basic emitter-follower voltage–source configuration; V_{ref} set by resistor ratio.

limitation and produces a reference voltage equal to the zener-diode breakdown voltage. The approximate output impedance of the basic circuit of Figure 3.12 is given by

$$R_{out} \approx \frac{1}{\Lambda I_1} + \frac{R_1 \parallel R_2}{\beta} \tag{3.31}$$

where I_1 is the load current and β is the current gain of the emitter follower. (The vertical bars denote the parallel combination of R_1 and R_2). The output impedance for the modified zener-diode circuit is

$$R_{out} \approx \frac{1}{\Lambda I_1} + \frac{\left[R_Z + \frac{R_1}{\Lambda(V_{CC} - V_Z)} \right] \parallel R_1}{\beta} \tag{3.32}$$

where R_Z is the dynamic resistance of the zener diode and V_Z is the zener-diode potential. Using typical device and circuit parameters, the range of practical output impedances reflected in Equations 3.31 and 3.32 is between 10 and 300 Ω. The term $1/\Lambda I_1$ is the output impedance of the basic emitter-follower stage and is the dominant term in both equations. Since this term is inversely proportional to the load current I_1 some compensation does occur for heavier loading.

We will now examine a biasing circuit, capable of producing a very stable reference voltage in the presence of temperature variations. This circuit, called a balanced-bridge configuration, is shown in Figure 3.13. The bridge is balanced for the conditions of

$$R_1 = R_2 \qquad \text{and} \qquad I_1 R_1 = n V_D \tag{3.33}$$

where V_D is the junction potential of one of the matched diodes in the diode strings. For the balanced condition, the current through the center leg composed of resistors R_A and R_B is negligible, and the reference voltage is simply $n V_D$. Neglecting thermal effects in D_Z, the temperature coefficient of V_{ref} does depend on the ratio of R_A and R_B and can be expressed as

$$\frac{\partial V_{ref}}{\partial T} = n \frac{\partial V_D}{\partial T} \frac{R_1 R_B}{R_A + R_B} \left[1 - \frac{R_A R_1}{R_B(R_1 + 2R_0)} \right] \tag{3.34}$$

By setting R_A and R_B resistor ratios according to

$$\frac{R_B}{R_A} = \frac{R_1}{R_1 + 2R_0} \tag{3.35}$$

the temperature coefficient of V_{ref} can be reduced to zero independent of the number of diodes used.

As we have seen, the technique of setting up a balanced-bridge network

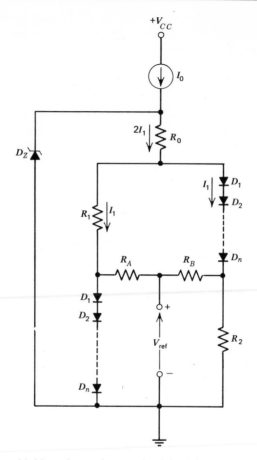

Figure 3.13   Balanced-bridge voltage reference circuit featuring good temperature stability.

with matched currents in each leg is a useful principle for differential amplifiers, biasing circuits, and current mirrors. We now demonstrate how all these basic circuits can be incorporated to form a differential amplifier circuit, biased with active devices acting as current-mirror loads.

3.5   CURRENT MIRRORS AND THE DIFFERENTIAL AMPLIFIER

Frequently the need arises in analog IC designs to produce a current signal that varies linearly with the applied differential input voltage. Current–signal sources have high impedance levels. As such, the typically

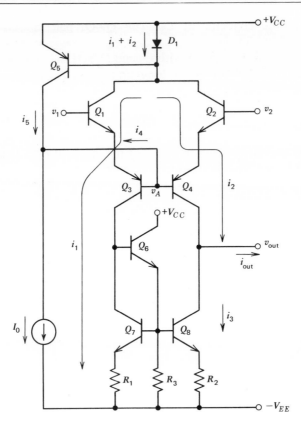

Figure 3.14 Differential input stage illustrating the current-mirror technique. (From *Analog Integrated Circuit Design* by A. Grebene, © 1972 by Litton Educational Publishing, Inc. Reprinted by permission of Van Nostrand Reinhold Company.)

small values of integrated capacitance can be connected at these circuit points to internally compensate the analog IC for frequency stability.

Figure 3.14 provides a circuit example of the current-mirror principle and illustrates the incorporation of several basic circuit configurations previously considered. The combined Q_1–Q_3 and Q_2–Q_4 pairs form an NPN–PNP differential cascade stage with Q_1 and Q_2 in the common-collector configuration and Q_3 and Q_4 in the common-base configuration. Matched transistors Q_7 and Q_8 act as high-impedance loads on this differential-input stage. The Q_6 transistor sets the bias level for Q_7 and Q_8 as well as serving as an emitter-follower stage to convert the collector-voltage of Q_7 into base-voltage drive for Q_8. Thus a single-ended output is available at the collector of Q_8. The Q_3–Q_4 pair is biased by the constant-current sink I_0.

The current sink I_0 together with Q_5 and D_1 sets the biasing currents of i_1 and i_2 for the amplifier. Again assuming geometric match between Q_5 and D_1, we can write

$$i_1 + i_2 = i_5 \left(1 + \frac{1}{\beta_p}\right) \tag{3.36}$$

where β_p is the current gain of either of the matched PNP transistors Q_3 and Q_4. The total base current of this pair is

$$i_4 = \frac{i_1 + i_2}{\beta_p} \tag{3.37}$$

Recognizing that

$$I_0 = i_4 + i_5 \tag{3.38}$$

the previous three equations simplify to show the desired current mirror effect of

$$I_0 = i_1 + i_2 \tag{3.39}$$

Thus we see that the operating currents of the stage are set by I_0 regardless of β_p. Therefore, the transconductance of the differential stage is well behaved, being set directly by I_0.

Currents i_1 and i_2 form a second current-mirror pair. Close device matching is necessary between the following transistor pairs: Q_1 and Q_2, Q_3 and Q_4, Q_7 and Q_8, and Q_5 and D_1. With no differential signal voltage applied ($v_1 = v_2 = V_{const}$), mirror currents i_1 and i_2 will be equal, assuming no offset effects. For this condition, $i_2 = i_3$ and the differential output-signal current, i_{out} is equal to zero. Now consider that v_1 is kept equal to V_{const} while v_2 is increased by ΔV_2 ($\Delta V_2 > 0$) to a new voltage of $V_{const} + \Delta V_2$. Thus the applied differential signal is ΔV_2. Because of the following action of the four base–emitter junctions, the voltage v_A will increase from its previous value by $\Delta V_2/2$. The effective driving voltage that controls the i_1 current, namely, $v_1 - v_A$, decreases because of the $\Delta V_2/2$ increase of v_A. Therefore, the i_1 current will also decrease proportionately. A reverse effect occurs for the i_2 current because its driving voltage increases by $\Delta V_2/2$. Therefore, i_2 increases by exactly the same amount that i_1 decreases.

Now consider the current levels in the Q_7 and Q_8 transistors. Through proper device matching and with $R_1 = R_2$, the i_1 and i_3 currents are directly proportional to each other by the ratio of their emitter areas. If we assume, for simplicity, equal areas, $i_1 = i_3$ for all conditions. Under the influence of the ΔV_2 signal voltage, i_3 will decrease because of the decrease in i_1. When coupled with the i_2 increase, this i_3 decrease produces a net increase in the output-signal current i_{out} (see Figure 3.14). Thus we see that the current-mirror effect between i_1 and i_3 produces a single-ended, output-current signal linearly proportional to the applied differential input voltage. For this reason, the entire circuit of Figure 3.14 is frequently considered to be a differential transconductance amplifier and modeled simply as in Figure 3.15.

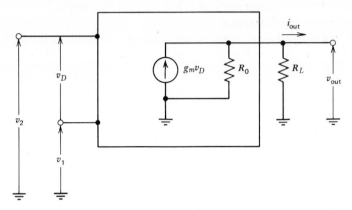

Figure 3.15 Simplified differential input transconductance amplifier model for circuit of Figure 3.14.

A final point should be noted concerning the circuit of Figure 3.14. The PNP transistors used in this circuit are usually lateral types whose characteristics are not as well controlled as those for the NPN types. In addition to setting the $i_1 + i_2$ current level, Q_5 and D_1 provide some stabilizing, common-mode feedback to the Q_1–Q_2 current mirror. The D_1 diode monitors the current level through Q_1 and Q_2. Because of Q_5–D_1 matching, Q_5 corrects for common-mode fluctuations in currents i_1 and i_2. For example, suppose an external influence causes both currents to increase simultaneously. This increase in the i_1 and i_2 currents would produce an increase in the Q_3–Q_4 bias current i_4. A decrease in i_5 is the result of an increase in i_4, since I_0 is a constant-current sink. The D_1 diode current ($i_1 + i_2$), being matched to i_5, must decrease. Hence common-mode negative feedback is provided by Q_5, I_0, and D_1 to stabilize the lateral PNP devices. The feedback is identified as common mode because the same effect is produced on both i_1 and i_2. This type of negative feedback simultaneously improves the bias stability and the common-mode rejection ratio without affecting the differential gain characteristics of the circuit.

3.6 EMITTER FOLLOWERS

The emitter-follower circuit is a well-established basic stage configuration that has found wide usage in both discrete and integrated circuit designs. The operation and complete circuit analysis of the emitter-follower, or common-collector stage as it is more precisely identified, is well documented in the literature. Figure 3.16 diagrams the basic emitter-follower circuit together with circuit equations describing its small-signal, low-frequency performance.

The emitter-follower stage is used almost entirely to provide signal isolation, since the configuration can exhibit a very high input impedance

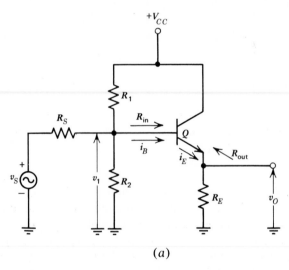

(a)

Circuit Parameter	Approximate Expression
Current gain, i_E/i_B	$h_{fe} + 1$
Input resistance, R_{in}	$h_{ie} + (h_{fe} + 1)R_E$
Voltage gain, v_O/v_1	$\dfrac{(h_{fe} + 1)R_E}{h_{ie} + (h_{fe} + 1)R_E} \approx 1$
Output resistance, R_{out}	$\dfrac{h_{ie} + R_1 \parallel R_2 \parallel R_S}{h_{fe} + 1}$

(b)

Figure 3.16 The emitter follower. (a) Basic circuit configuration. (b) Expressions describing ac performance.

and a very low output impedance. Thus emitter followers do not require as much signal current from the preceding stage as some other stage configurations. Because of the follower's low output impedance, this stage is capable of supplying a large amount of load current to subsequent stages.

Most of the desirable characteristics of an emitter follower arise because of its inherent negative feedback. The base–emitter junction of Q acts as a summing point between the applied signal voltage v_i and the entire output voltage v_o. Even though the voltage gain of the stage is always slightly less than unity, the current gain is approximately equal to that of the transistor; thus appreciable power gain is possible with this circuit. Emitter followers are used principally in the biasing circuit structures as the reference voltage sources we have previously examined and in the output stages of AICs to provide a large signal-drive capability. We examine several output-stage configurations later in this chapter.

3.7 GAIN STAGES

Gain stages employed in analog ICs are usually intermediate stages located between the input-circuit configuration and the output stage. Because of the isolation from external circuit influences offered by the intermediate location, the design criteria for a gain stage are less severe than for either the input or the output stages, which must interface with signal sources and various loading conditions.

As its name suggests, a gain stage serves mainly to boost the signal level from the input stage to an amplitude satisfactory for driving the output stage. The particular type of signal gain supplied by this intermediate stage can take several forms, depending on the impedance characteristics of the input and output stages. For example, if the output impedance of the input stage is very small (as provided from an emitter-follower stage) and if the input impedance of the output stage is very large (again such as provided by an emitter follower), a gain stage producing voltage gain would be most appropriate. On the other hand, if impedance levels were reversed between the input and output stages, a current-gain stage would be the most appropriate choice. Figure 3.17 tabulates the input and output impedance conditions for the four common types of gain stage. Frequently the various gains produced by this stage constitute most of the high overall gain achieved in analog IC devices such as operational amplifiers and current boosters.

Gain stages have an additional function to internally shift the quiescent voltage level developed in the input stage. The voltage must be shifted to

Gain Stage Type	Output Impedance of Input Stage	Input Impedance of Output Stage
Voltage gain	Low	High
Current gain	High	Low
Transconductance gain	Low	Low
Transimpedance gain	High	High

Figure 3.17   Impedance considerations in typing intermediate-gain stages.

a level that will enable the output stage to properly interface with its intended load. For example, in operational amplifiers the gain stage would provide a dc level shift so that the quiescent output level is zero volts. Gain stages utilized in voltage comparators are designed to interface with the output stage in such a way that the output voltage levels are compatible with a particular logic type.

Isolated from external circuit influences by virtue of their intermediate position, gain stages are a logical place to provide various types of circuit compensation. Adjustment of the output offset voltage in analog ICs is frequently accomplished in the gain stage, as well as frequency compensation for stabilizing operational amplifiers. We investigate these characteristics in more detail as we study the operation of particular analog IC devices.

An ever-increasing variety of gain-stage configurations has been designed to meet specialized design criteria. The preponderance of such circuits makes it difficult to define a basic gain stage. However, we can illustrate several types that have found wide acceptance and are reasonably indicative of the basic design philosophies frequently incorporated into this stage.

The first gain stage we examine is the circuit of Figure 3.18, which is a basic differential amplifier utilizing PNP transistors. This PNP gain stage is being driven by a similar NPN differential amplifier input stage. A single-ended output voltage taken from the gain stage drives the emitter follower to buffer the output stage.

Many analog IC designs use the NPN–PNP differential amplifier cascade to achieve voltage gain and level shifting. The dc bias level developed in the first differential stage is essentially canceled by the gain stage and emitter follower. Level shifting is provided in the gain stage by using transistors opposite in conductivity type to those of the input stage; the base–collector junctions provide the cancellation effect.

The current mirror offers a convenient circuit position to effect an

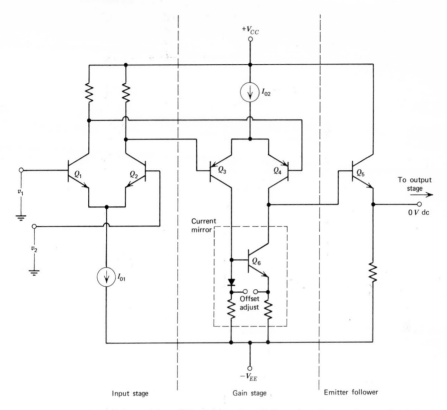

Figure 3.18 PNP differential amplifier with emitter follower used as a voltage gain stage.

adjustment of the output-offset voltage. Alteration of the current gain of the mirror can be used as a means of adjusting the quiescent collector voltage of Q_4. Since the emitter follower provides a near-constant level shift of one forward-biased junction potential, adjustment of the collector voltage of Q_4 will be transmitted directly to the output stage.

Compensation techniques to reduce the tendency of high-frequency oscillations usually are accomplished in the gain stage. A small capacitor connected between the collectors of Q_3 and Q_4 will reduce the high-frequency gain of the stage, thereby lessening the circuit's tendency to oscillate. A capacitive network inserted between the base of Q_3 and the collector of Q_4 is an alternative approach for reducing the high-frequency gain of this stage.

A second example of a gain stage, structured with the basic Darlington Q_1–Q_2 connection, appears in Figure 3.19. Voltage gain and level shifting

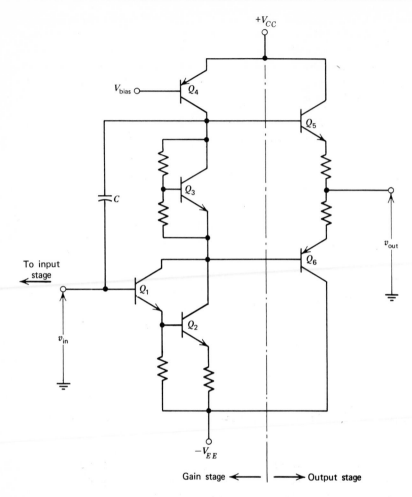

Figure 3.19   Gain stage using the Darlington configuration to drive a complementary output stage.

for positive voltages are accomplished in this stage. Transistor Q_3 provides an additional positive-level shift to drive the NPN transistor Q_5—one-half of the NPN–PNP complementary output stage. Capacitor C, which provides frequency compensation, reduces the high-frequency gain by shunting these frequencies around the gain stage. Because the impedance levels of both capacitor terminals are extremely large (collectors and bases of emitter-follower stages), a small value of capacitance can be employed to achieve a relatively low break frequency. This is a particular advantage in many analog IC designs in which the capacitor C is fabricated onto the monolithic substrate.

A more sophisticated example of a gain stage (Figure 3.20), an intermediate amplifier stage, has a very low input impedance because of the common-base configuration of the Q_1–Q_2 differential pair and the very large output impedance at the collectors of Q_3 and Q_4. Hence this stage provides a large voltage gain as well as a downward-level shifting by the

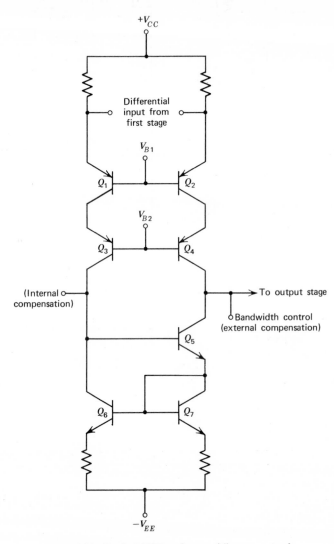

Figure 3.20 Basic gain stage for providing current gain.

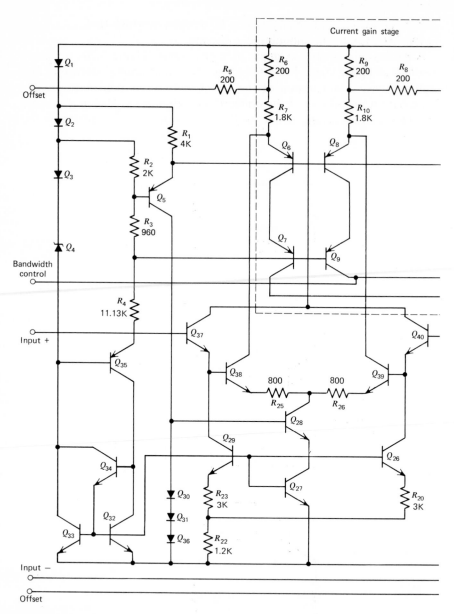

Figure 3.21 Circuit schematic of HA-2500 operational amplifier, outlining basic current value is kilohms.

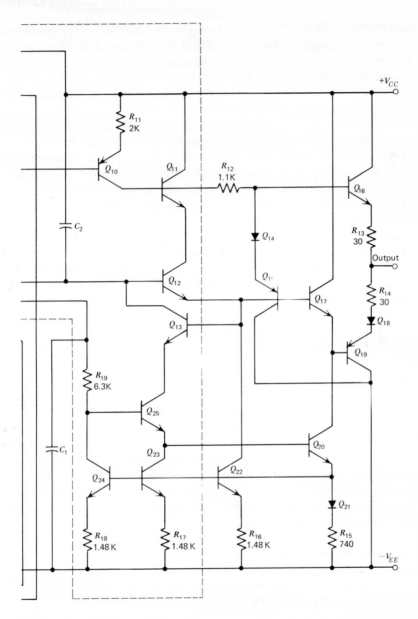

gain stage. If only number appears, it is measured in ohms; if K appears with number, its

four PNP transistors. Transistors Q_5–Q_7 form a current mirror, providing differential-to-single-ended conversion at the collector of Q_5. If equal and opposite currents ΔI are applied to the emitters of Q_1 and Q_2, the current change at the output of the stage will be $2\Delta I$; consequently, the voltage gain of the stage is the product of the input-stage transconductance and the impedance at the collector of Q_4.

The high impedance present at the collector of Q_4 makes this point ideal for bandwidth control, since the pole affected here can be made the dominant pole of the amplifier with a small-value capacitor. The collector of Q_3 is frequently used as an internal compensation point to generate a lag-lead type of response near the bandwidth of the amplifier to improve phase margin. The collector of Q_4 is normally brought out to an external terminal for external compensation.

All PNP transistors should be high-frequency devices for improved frequency response. Good device characteristics coupled with the common-base configuration ensure near-optimum frequency response from this basic gain stage. Figure 3.21 incorporates this gain stage in an operational amplifier.

3.8 OUTPUT STAGES

The output stage of an analog IC device is designed to deliver a substantial amount of power into a low-impedance load. The characteristics of output stages vary because not all analog ICs are designed to meet the same specifications. For example, some designs require single-ended outputs, whereas others specify a differential output. In some applications where harmonic distortion is a major consideration, the output stage would probably be designed for class A operation. On the other hand, the need to achieve high conversion efficiency may require the output stage to function class AB, or even class B. In general, the output stage should have the following properties:

1. Low output impedance
2. Large output voltage and current swing capability.
3. Low standby power dissipation.
4. Short-circuit protection.

The emitter follower discussed in Section 3.6, which exhibits a low output impedance and large voltage and current capabilities, can easily be protected against short circuits by a simple circuit modification. However, a single transistor used in the emitter-follower configuration is capable of

driving loads effectively in only one direction: the direction (positive or
negative) depending on transistor type. Furthermore, the emitter-follower
configuration requires an amount of standby power approximately equal
to the load power. Figure 3.22 displays several modifications commonly
made to the basic emitter-follower configuration to develop a basic output

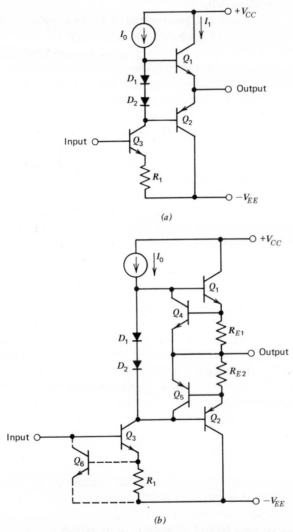

Figure 3.22 A class AB output stage. (*a*) Basic configuration. (*b*) Modification for
short-circuit protection. (From *Analog Integrated Circuit Design*, by A. Grebene, © 1972 by
Litton Educational Publishing, Inc. Reprinted by permission of Van Nostrand Reinhold
Company.)

stage that circumvents the shortcomings cited. Device matching exists between D_1 and Q_1 and between D_2 and Q_2. For no applied input signal to the circuit in Figure 3.22a, the quiescent current in the both transistors I_1 is proportional to the constant-current source I_0 and to the ratio of emitter areas. Thus both Q_1 and Q_2 can be barely active in the standby condition and can operate class AB.

A positive-input signal will be inverted by Q_3 and applied to the base of Q_2. Transistor Q_2 would probably be fabricated as a substrate PNP device, to have a high current capabilitiy for sinking load current to the $-V_{EE}$ supply. Negative input signals into Q_3 will be inverted and will turn Q_2 off and drive Q_1 into heavy saturation. Thus Q_1 can act as a source of load current from the V_{CC} supply for negative inputs and Q_2 can act as a sink of load current to the $-V_{EE}$ supply for positive inputs.

An accidental short circuit of the output terminal to one of the power supplies will probably destroy one of the Q_1 or Q_2 transistors because of the drawing of excessive current. The easiest modification for achieving short-circuit protection is to insert series resistors between the emitters and the output terminal. These resistors will limit the maximum device current to approximately V_{CC}/R_{E1} or V_{EE}/R_{E2} where R_{E1} and R_{E2} are the emitter resistors associated respectively with Q_1 and Q_2. To ensure perfectly safe operation, however, it may be necessary to use excessively large values of emitter resistors. Such large resistors will increase the output impedance of the output stage, thereby seriously limiting its output drive capability.

The further modification to the emitter-follower configuration of Figure 3.22b solves the need for excessively large series resistors. Under normal operation, Q_4 and Q_5 are turned off and do not affect the circuit's operation. However under short-circuit conditions, as the output current increases, sufficient voltage can develop across R_{E1} or R_{E2} to turn the corresponding transistor on. For example, before Q_1 can develop excessive current, Q_4 will turn on and shunt base-current drive from Q_1. This action causes the output current to limit at an approximate value of

$$I_{max} \approx \frac{V_{BE}}{R_E} \tag{3.40}$$

This scheme allows an order-of-magnitude reduction in the size of R_E needed to protect the output transistors against short circuits. Typical values of I_{max} range between 10 and 100 mA.

It may be necessary to add an additional transistor Q_6 to give transistor Q_3 protection from short-circuits. Under the influence of a large input signal, Q_3 and Q_5 establish a current path between $-V_{EE}$ and the output; Q_6 will limit the current in this path to a safe level.

With a modification of the circuit in Figure 3.22*b*, another basic output stage can be formed. Figure 3.23 shows the circuit that is developed by interchanging the positions of Q_3 and I_0. In the modified circuit, the impedance level at the base of Q_3 is increased because of the emitter-follower configuration of Q_3. Hence frequency compensation is often

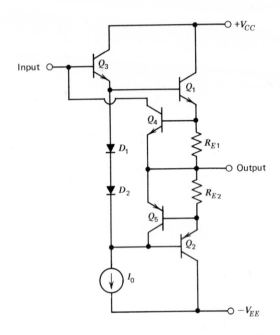

Figure 3.23   A basic output stage achieved by interchanging the positions of Q_3 and I_0 in Figure 3.22*b*. (From *Analog Integrated Circuit Design* by A. Grebene, © 1972 by Litton Educational Publishing, Inc. Reprinted by permission of Van Nostrand Reinhold Company.)

provided at this high-impedance point. This output stage provides unity voltage gain to the input signal without phase inversion. However the stage does produce a current gain of approximately h_{fe}^2 to the signal current applied at the base of Q_3.

A final basic output stage (Figure 3.24) employs paired NPN and PNP transistors in parallel paths between the input and output terminals. The operation of this circuit is quite similar to the previous configuration of Figure 3.23 with the exception of the two diodes. These diodes normally do not conduct. However, if a large-signal voltage appears at the input, one of the paired transistors in the parallel paths can become turned off. Under this condition one of the diodes will become forward-biased,

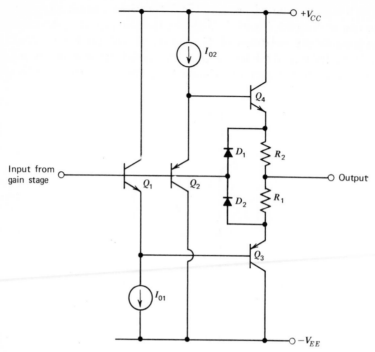

Figure 3.24 Basic output stage with buried NPN–PNP transistors in parallel circuit paths to output.

providing a third path between input and output. This basic output stage is employed in the HA-2600 (see Figure 3.25).

SPECIALIZED CIRCUIT STRUCTURES

The basic circuit configurations generally provide acceptable performance for most analog applications. The characteristics of these structures are well understood by circuit designers who use these configurations for constructing a complete integrated circuit

In a number of analog IC applications, however, the performance that is called for cannot be achieved with the basic circuit structures. Specialized circuit design techniques must then be employed, involving modification of the basic structures to achieve improved circuit characteristics. The following sections describe several of these specialized design techniques.

3.9 MODIFIED DIFFERENTIAL AMPLIFIER

The basic differential amplifier discussed in Section 3.1 has an input-voltage–output-current characteristic described by the hyperbolic tangent function as given in Equation 3.16. The output current is asymptotic to the bias current I_0 as $|v_D|$ becomes large (see Figure 3.6). Thus I_0 is the maximum differential current that can be generated by this basic amplifier.

In some applications it is desirable to have a differential amplifier capable of generating differential output currents much larger than the equilibrium bias-current level. In such an application, for example, the differential amplifier might be used as the input stage in a low-power, high-speed operational amplifier. A circuit that does not have the output current limitation appears in Figure 3.26a. In this special circuit, the NPN transistors are a matched pair and the four PNP devices are also closely matched. Using a circuit analysis approach similar to that presented in Section 3.1, it can be shown that the differential output current is related to the differential input voltage as

$$i_D = 2I_0 \sinh \frac{\Lambda v_D}{2} \tag{3.41}$$

The transfer characteristic described by this equation is graphed in Figure 3.26b, which shows no output-current limitation. Of course practical device constraints will limit the differential output current to finite levels. However, this specialized circuit configuration is capable of producing output currents several hundred times as large as the equilibrium bias-current level.

3.10 FET CIRCUITS

FETs have certain inherent characteristics that can be exploited in specialized configurations to improve circuit performance. The extremely high impedance levels at the gate of the FET and the very low channel resistance with no offset voltage are best utilized in applications such as source followers and analog switches.* (FETs are not commonly used in gain stages because of their relatively low g_m.)

FETs acting as source followers in the input stage have a distinctive advantage over bipolar devices because FETs require exceptionally low input-bias current. Such low currents are achieved at the expense of

---

* Analog switches with FETs are considered in Section 3.15.

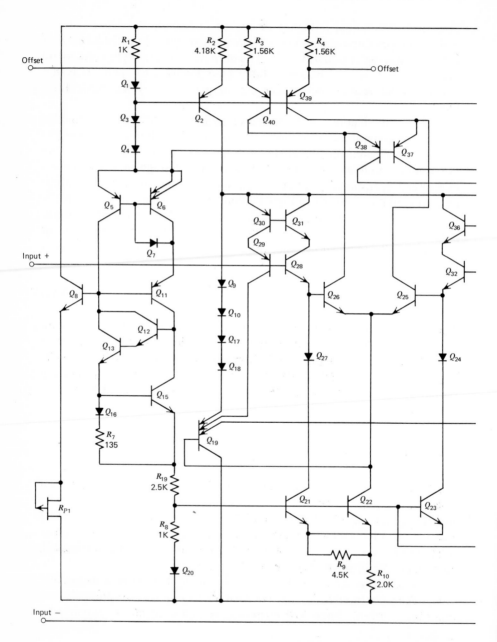

Figure 3.25 Complete circuit diagram

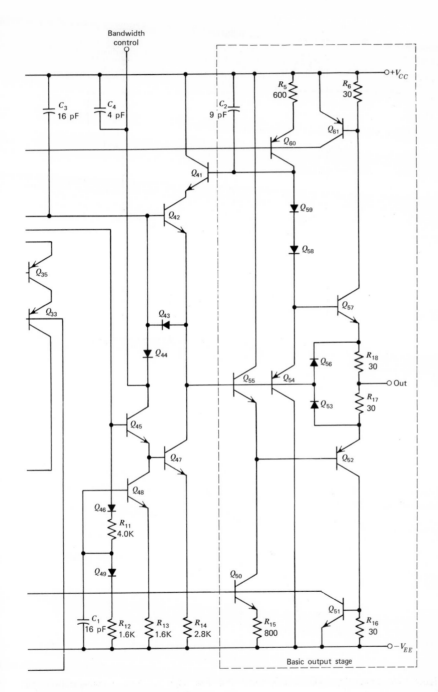

of the HA-2600 operational amplifier.

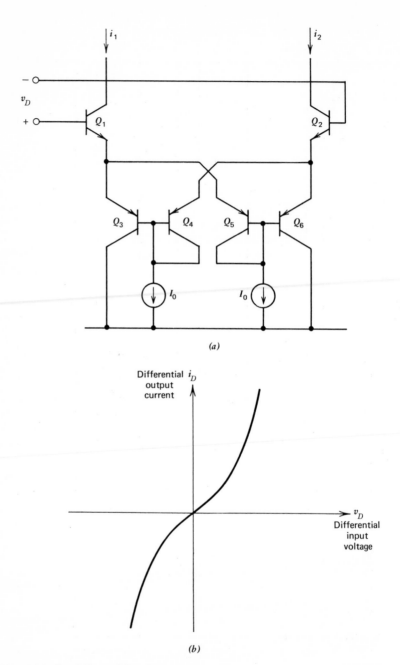

(a)

(b)

Figure 3.26 Specially modified differential amplifier stage. (a) Circuit configuration. (b) Transfer characteristic.

increased noise levels and larger input-offset voltages than can be obtained with conventional bipolar front ends. Generally speaking, the specific application of the analog IC determines the tradeoffs that can be made.

As a rule, JFETs have less white noise than MOS FETs and are preferred as components in the input stage unless extremely low currents at high temperatures are required. Figure 3.27 diagrams a complete input stage in which JFETs are the input devices. This circuit achieves an extremely low maximum bias current of 20 pA at 25°C and a maximum bias current of 10 nA over the temperature range of −55 to +125°C.

The preamplifier is specifically designed to interface with operational amplifiers to form hybrid FET amplifiers whose frequency characteristics are defined by the op amps themselves and are not limited by the FETs. The three JFETs are matched devices and F_{1M} sets the Q-point of the input devices. The FETs operate in the saturated mode where the gate-to-source voltage V_{GS} maximizes individual FET g_m and ensures close g_m matching among all FETs. The saturated mode also minimizes offset voltage because of biasing-current mismatch. However, setting V_{GS} exactly equal to zero is not practical because of expected V_{GS} mismatches. A nominal V_{GS} of 0.1 V is intentionally introduced by altering the geometry of F_{1M} with respect to F_{1AM} and F_{1BM}. All three FETs are constrained to carry the same current because of the current mirror of Q_{10S}, Q_{9M}, Q_{9AM}, Q_{9BM}, and the R resistors. Operating with $V_{GS} = 0.1$ V achieves good common-mode range for the composite preamplifier because the front end does not limit the input specification of the associated operational amplifier.

Transistors Q_{8AS}, Q_{8BS}, Q_{5AS}, and Q_{5BS} form two pairs of stacked constant-current circuits (discussed in the following section) which reduce h_{rb} effects in the NPN and PNP transistors, thereby minimizing supply-current sensitivity to power-supply variations and common-mode input-voltage variations. The contribution of the FET front end to common-mode and power-supply rejection ratios for the composite amplifier is negligible.

Transistors Q_{1AM} and Q_{1BM} provide an output current capability to charge the effective input capacitance of the op amp thereby relieving slew-rate limitations. The FETs are bootstrapped to circumvent their inherently low gate-to-drain breakdown voltage of approximately 8 V. Pinched resistor F_2 provides a turn-on current path through Q_{13S} and the four diodes D_5 through D_8 to produce the desired gate-to-drain voltage across F_{1M}. Capacitor C is an internal element used in the hybrid application to neutralize the pole produced by the feedback resistor and the input capacitance of the operational amplifier.

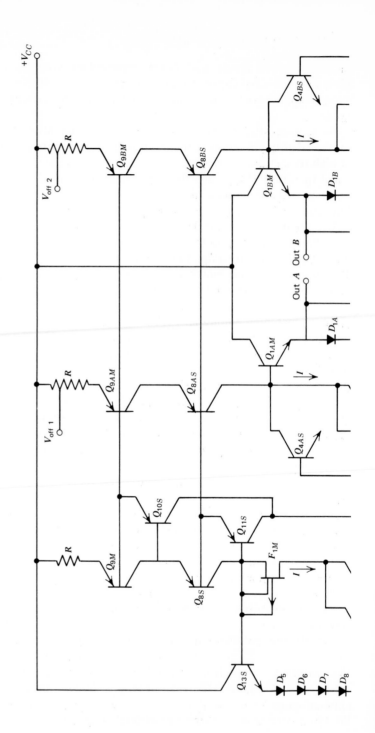

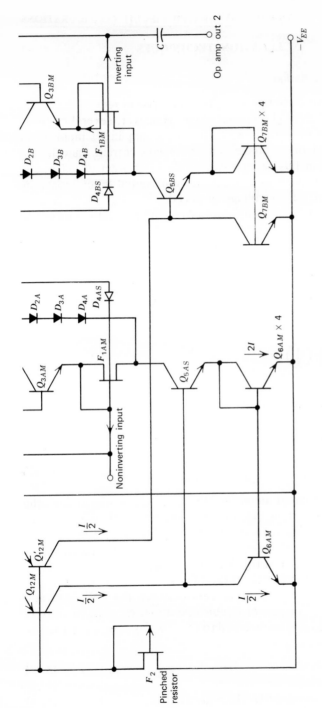

Figure 3.27 FET preamplifier (HA-2000).

3.11 ADMITTANCE CANCELLATION TECHNIQUES

Base-Current Cancellation

Several special circuit techniques can be employed to reduce the input-bias current in a differential amplifier stage. However, several of these techniques introduce intolerable tradeoffs in the amplifier's input-offset voltage and/or speed of response. The base bias-current cancellation technique discussed in this section can produce an order-of-magnitude improvement in the input-bias current and differential input resistance without compromising offset voltage or speed.

Figure 3.28 shows an NPN transistor biased with an emitter current I_0.

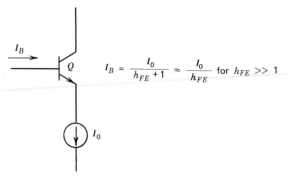

$$I_B = \frac{I_0}{h_{FE}+1} \approx \frac{I_0}{h_{FE}} \text{ for } h_{FE} \gg 1$$

Figure 3.28   Current-source biasing technique.

The base current for Q is

$$I_B = \frac{I_0}{h_{FE}+1} \approx \frac{I_0}{h_{FE}} \tag{3.42}$$

for $h_{FE} \gg 1$. In a differential amplifier, the base current determines the input-bias and offset-current parameters. The most obvious way of reducing I_B is by reducing I_0 or increasing h_{FE}. Reducing I_0 lowers the stage gain and reduces the slew rate. Super-β transistors can increase h_{FE}, but at the expense of reduced breakdown voltages.

The circuit in Figure 3.29 increases the effective h_{FE} of the Q_1 transistor by a cancellation of base-biasing currents between the PNP (Q_4) and NPN (Q_1) transistors. For proper circuit operation, Q_1 must be closely matched to Q_2, and Q_3 must be matched to Q_4. If we assume $h_{FE} \gg 1$ for all transistors, we can write

$$I_0 \approx I_{E2} \approx h_{FE2}I_{B2} \tag{3.43}$$

and

$$I_{E4} = h_{FE3}I_{B2} \approx h_{FE4}I_{B4} \tag{3.44}$$

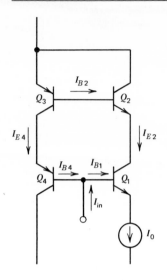

Figure 3.29 Circuit for base biasing-current cancellation.

Also, we have

$$I_{in} + I_{B4} = I_{B1} \qquad (3.45)$$

After some algebraic reduction, we have the desired result

$$I_{in} \approx \frac{I_0}{h_{FE1}}\left(1 - \frac{h_{FE1}h_{FE3}}{h_{FE2}h_{FE4}}\right) \qquad (3.46)$$

Thus we see that this cancellation technique increases the effective h_{FE} of a transistor by a factor of

$$\left(1 - \frac{h_{FE1}h_{FE3}}{h_{FE2}h_{FE4}}\right)^{-1}$$

If the $h_{FE} \gg 1$ assumption is removed, the exact expression for I_{in} is

$$I_{in} = \frac{I_0}{h_{FE1}+1}\left(1 - \frac{h_{FE1}h_{FE3}}{(h_{FE2}+1)(h_{FE4}+1)}\right) \qquad (3.47)$$

Successful implementation of this cancellation technique requires close device matching between Q_1 and Q_3 and between Q_2 and Q_4. By rearranging the $(h_{FE2}+1)$ and $(h_{FE4}+1)$ terms, we can approximate Equation 3.47 as

$$I_{in} \approx \frac{I_0}{h_{FE1}+1}\left[1 - \left(1+\frac{\Delta h_{FEN}}{h_{FE2}}\right)\left(1-\frac{1}{h_{FE2}}\right)\left(1+\frac{\Delta h_{FEP}}{h_{FE4}}\right)\left(1-\frac{1}{h_{FE4}}\right)\right] \qquad (3.48)$$

where Δh_{FEN} and Δh_{FEP} represent the respective mismatches between the NPN and PNP transistor pairs. For typical $h_{FE} \geqslant 100$ and matching

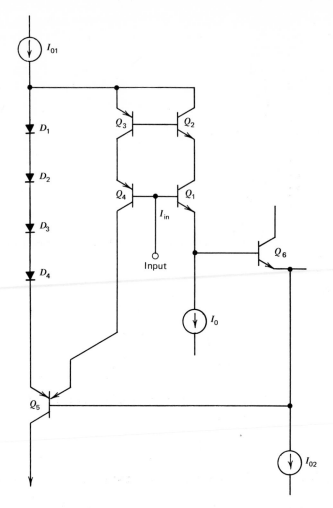

Figure 3.30 Complete circuit for input base-current cancellation.

between transistors of

$$\frac{\Delta h_{FE}}{h_{FE}} \leqslant 5\% \qquad (3.49)$$

the input current becomes

$$I_{in} \leqslant \frac{I_0}{10(h_{FE1}+1)} \qquad (3.50)$$

which shows an order-of-magnitude reduction in the input current. It can be demonstrated that the same improvement factor is achieved in the differential input resistance.

Usually some additional circuitry is necessary to realize optimum transistor h_{FE} matching. In the complete circuit presented in Figure 3.30, the four diodes ensure that all matched transistors are biased at equal collector-to-base voltages. Furthermore, the bootstrapping provided by the dual-emitter transistor Q_5 and the four diodes minimizes the input capacitive effects.

Neutralization

Many analog IC applications require the establishment of extremely large output resistances ($>10\ \mathrm{M}\Omega$) for constant-current circuits and other functional mechanizations. Achieving these very large resistance levels (small conductance levels) calls for specialized circuit design techniques to reduce the dominant influences of the h_{rb} and h_{ob} parameters. In this section, we discuss some neutralization techniques for reducing these parameters.

h_{rb} *Neutralization*　Our examination of the basic current-mirror circuit of Figure 3.10b revealed that the h_{rb} term dominated the output conductance of this circuit. An efficient solution for reducing the h_{rb} influence is to utilize a stacked current-mirror configuration (see Figure 3.31a). The stacking of the additional D_1–Q_3 combination forces Q_2 to operate at approximately $V_{BC} = 0$ V.

The presence of Q_3 reduces considerably the effects of voltage feedback because of the h_{rb} parameter in the hybrid model. In the stacked circuit, the h_{ob} term in Equation 3.25 remains essentially unchanged because the additional resistance of D_1 is negligible with respect to $(h_{ob})^{-1}$. Furthermore, the negative-feedback loop composed of D_1, Q_2, and Q_3 further stabilizes the I_0 current. The output conductance of the stacked circuit reflects the neutralization of the h_{rb} component because the impedance at the emitter of Q_3 looking into the collector of Q_2 is $(g_{\text{out}-Q2})^{-1}$. Therefore, the output conductance at the collector of Q_3 can be found as

$$g_{\text{out}-Q3} \approx h_{ob3}\left(1 + \frac{h_{fb2}h_{fb3}h_{ib1}}{h_{ib2}}\right) - \frac{h_{fb3}h_{rb3}}{h_{ib3} - h_{ib2}/h_{fb2}h_{rb2}} \qquad (3.51)$$

Since $h_{fb} \approx -1$ and Q_1 and Q_2 are matched, we can approximate Equation

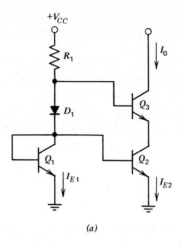

$+V_{CC}$

R_1

I_0

Q_3

D_1

Q_1

I_{E1}

Q_2

I_{E2}

(a)

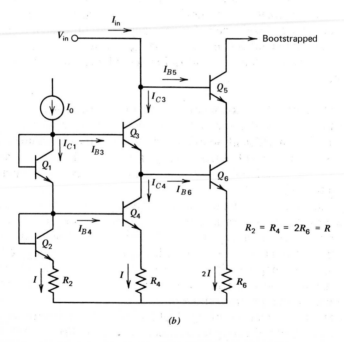

I_{in}

V_{in}

Bootstrapped

I_{B5}

I_{C3}

Q_5

I_0

Q_3

I_{C1} I_{B3}

Q_1

I_{C4} I_{B6}

Q_6

Q_4

I_{B4}

$R_2 = R_4 = 2R_6 = R$

Q_2

I R_2

I R_4

$2I$ R_6

(b)

Figure 3.31 Neutralization circuits. (a) h_{rb} neutralization using the stacked current mirror. (b) h_{ob} neutralization circuit.

3.51 by

$$g_{out-Q3} \approx 2h_{ob3} + \frac{h_{rb3}h_{rb2}}{h_{ib2}} \approx 2h_{ob3} \qquad (3.52)$$

Thus the h_{rb} effect of Q_3 is essentially neutralized by h_{rb2} of the stacked transistor, Q_2.

h_{ob} *Neutralization* When output resistances larger than the $(2h_{ob})^{-1}$ given by Equation 3.52 are required, we need more sophisticated circuit design techniques. Figure 3.31b illustrates a circuit that combines the previously discussed stacking technique for reducing h_{rb} together with a neutralization technique for reducing h_{ob} effects.

In Figure 3.31b, the Q_1, Q_2, Q_3, Q_4 combination is the stacked current mirror for h_{rb} neutralization. Hence the dominant input-conductance term to the stacked current mirror alone will be due to h_{ob} effects. A summing node occurs at the base of Q_6, where the conductance h_{ob} is fed back through Q_3 to effectively cancel the h_{ob} conductance between the base and collector of Q_3. In the circuit Q_2 and Q_4 are matched devices carrying equal emitter currents; the same is true for Q_1 and Q_3. Transistors Q_5 and Q_6 are fabricated with emitter areas twice as large as those of Q_2 and Q_4. Note that

$$R_2 = R_4 = 2R_6 \qquad (3.53)$$

Hence the emitter currents of all transistors are related as

$$I_{E5} \approx I_{E6} \approx 2I_{E1} \approx 2I_{E2} \approx 2I_{E3} \approx 2I_{E4} \qquad (3.54)$$

A first-order mathematical analysis is given in Appendix B to further illustrate the h_{ob} cancellation. This appendix demonstrates that

$$g_{in} = \frac{\Delta I_{in}}{\Delta V_{in}} \approx 2h_{ob3} - h_{ob6} \qquad (3.55)$$

Since Q_6 conducts twice the collector current of Q_3 and h_{ob} is directly proportional to the collector current, we have

$$h_{ob6} = 2h_{ob3} \qquad (3.56)$$

This equation implies that g_{in} approaches zero, or R_{in} approaches a true open circuit.

In addition to producing a virtually infinite impedance level at the collector of Q_3, the circuit in Figure 3.31b introduces no first-order differential current errors at the inputs where I_{in} and I_0 currents exist. For this reason, this h_{ob} neutralization technique is an extremely useful circuit for converter applications in which a differential signal is converted to a single-ended output. This circuit circumvents the common problem encountered in most amplifier designs of differential current errors, producing intolerable contributions to the input-offset voltage in the presence of temperature changes.

Practical limitations to the h_{ob} neutralization technique are due to second-order terms. When these are considered, the total output conductance for the circuit of Figure 3.31b can be described by

$$g_{out} \approx (2h_{ob3} - h_{ob6}) + \frac{h_{rb6}}{h_{fe5}(h_{ib6} + R/2)} + \frac{h_{rb3}h_{rb4}}{h_{ib4} + R} \qquad (3.57)$$

With the h_{ob} cancellation of Equation 3.56, g_{out} can be approximated by

$$g_{out} \approx \frac{h_{rb6}}{h_{fe5}(h_{ib6} + R/2)} + \frac{h_{rb3}h_{rb4}}{h_{ib4} + R} \qquad (3.58)$$

Bootstrapping As an alternative means of increasing the effective resistance at a circuit point, we have the voltage and/or current in a circuit "lift itself by its bootstrap" to produce the required driving capability. Bootstrapping is a form of positive feedback that is acceptable in analog ICs as long as the gain between the bootstrapped points is kept below unity. A gain greater than unity with positive feedback will cause instability.

Figure 3.32 illustrates a bootstrapping method to achieve neutralization of h_{rb} and h_{ob} contributions. With bootstrapping the voltage drops across forward-biased base–emitter and diode junctions are exactly equal to the voltage rises across similar junctions as one progresses around a loop. Bootstrapping exists between points A and B through two circuit paths that share a common branch. The combination Q_3, Q_4, Q_6, and Q_7 is a stacked current-mirror network that neutralizes h_{rb} of Q_6. The first bootstrapping path contains the Q_1, Q_2, D_3, Q_5, Q_4, and Q_7 devices and neutralizes the h_{ob} of Q_6. A second bootstrapping loop through Q_1, Q_8, D_2, D_1, and Q_9 neutralizes the h_{ob} of Q_1.

This bootstrapping technique achieves a near-zero single-ended conductance. At the collector terminal of Q_6, all first-order effects of h_{rb} and h_{ob} are neutralized to produce a circuit having a very low input-offset current and very high power-supply and common-mode rejection ratios. An approximate expression for the differential conductance is

$$g_{CQ6} \approx \frac{h_{rb3}h_{rb5}h_{rb6}}{h_{ib3}} + \frac{h_{ob5} + g_{i3}}{(h_{fe1}+1)(h_{fe2}+1)} + \frac{h_{ob7} + g_{i1}}{(h_{fe8}+1)(h_{fe1}+1)} + \frac{g_{i2}}{h_{fe1}+1} \qquad (3.59)$$

where g_{in} is the conductance associated with the nth constant-current circuit. Using typical parameters given in Chapter 2, $g_{CQ6} \approx 0.006\ \mu$mho for an equivalent resistance with respect to ac ground of 150 MΩ.

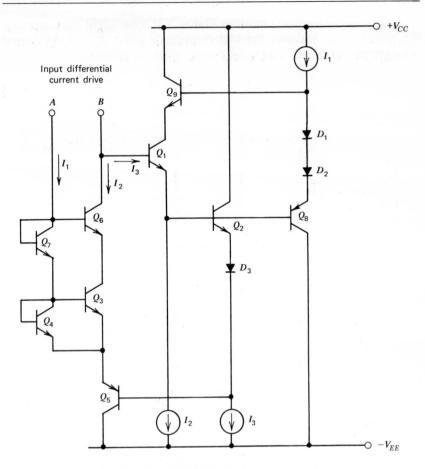

Figure 3.32 Bootstrapping circuit configuration.

3.12 POSITIVE-FEEDBACK NETWORKS

We frequently encounter special applications in which the power con-
sumption in an output stage must be kept at a minimum level while high
current gain must be provided to isolate the high-impedance point in a
circuit from the output terminal. One specialized circuit configuration (see
Figure 3.33) employs positive-feedback techniques to lower the quiescent
power dissipation while boosting the current gain of the input signal.

On first inspecting the circuit in Figure 3.33, one might postulate that the current gain between the high-impedance point A and the output through the cascaded emitter followers might be given by

$$A_I = \frac{i_{out}}{i_x} \approx (\beta_6 + 1)(\beta_9 + 1) \tag{3.60}$$

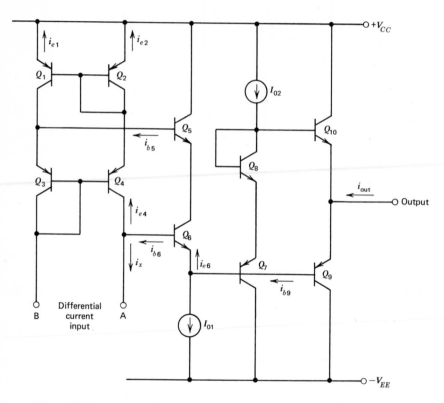

Figure 3.33 Low-power–high-current gain stage employing positive feedback.

This is indeed partially correct. However, the positive-feedback path provided by transistors Q_1 through Q_6 can make A_I significantly larger than the value predicted in Equation 3.60, as the following analysis indicates.

In the analysis, all currents are small-signal ac quantities, and close device matching exists between the following transistor pairs: Q_1 and Q_2, Q_3 and Q_4, Q_5 and Q_6, Q_8 and Q_{10}, and Q_7 and Q_9. From the circuit, we can write the following equations:

$$i_{b9} \approx i_{e6} \approx \frac{i_{\text{out}}}{\beta_9 + 1} \tag{3.61}$$

$$i_{b5} = \frac{i_{e5}}{\beta_5 + 1} = \frac{\alpha_6 i_{e6}}{\beta_5 + 1} \approx i_{e1} \approx i_{e2} \tag{3.62}$$

$$i_{e4} = (2 - \alpha_2)\alpha_4 i_{e2} \tag{3.63}$$

$$i_{b6} = \frac{i_{e6}}{\beta_6 + 1} \tag{3.64}$$

$$i_x = i_{b6} - i_{e4} \tag{3.65}$$

Combining Equations 3.61 through 3.65 gives

$$i_x = \frac{i_{\text{out}}}{\beta_9 + 1}\left[\frac{1}{\beta_6 + 1} - \frac{(2 - \alpha_2)\alpha_4\alpha_6}{\beta_s + 1}\right] \tag{3.66}$$

If the β's of Q_s and Q_6 are well matched, the preceding equation simplifies to

$$i_x = \frac{i_{\text{out}}}{(\beta_6 + 1)(\beta_9 + 1)}[1 - (2 - \alpha_2)\alpha_4\alpha_6] \tag{3.67}$$

and the current gain of the stage is

$$A_I = \frac{i_{\text{out}}}{i_x} = \frac{(\beta_6 + 1)(\beta_9 + 1)}{1 - (2 - \alpha_2)\alpha_4\alpha_6} \tag{3.68}$$

Thus the positive feedback increases the current gain of the stage by a factor of $[1 - (2 - \alpha_2)\alpha_4\alpha_6]^{-1}$.

3.13 ANALOG MIXING CIRCUITS

Certain analog IC applications, particularly in communications systems, require special configurations that obtain a single output signal as the product of two separate input signals. Circuits that perform this function are frequently called mixers. When the amplitude of the output signal is the primary parameter of interest, the mixer may be identified as a multiplier. When the phase difference between the two input signals, which appears at the output, is the principal parameter, the same mixer may be called a phase detector. In this section we discuss the operation of a special class of analog mixing circuits called transconductance amplifiers, which are particularly well suited for implementation in AICs.

The transconductance differential amplifier of Figure 3.5 is the basic circuit used to generate the analog mixer. In Equations 3.14 and 3.15 we showed that the collector currents in a differential amplifier are related to the difference input signal v_D as

$$i_1 = \frac{I_0 e^{\Lambda v_D}}{1 + e^{\Lambda v_D}} = I_0 F_1(v_D) \tag{3.69}$$

$$i_2 = \frac{I_0}{1 + e^{\Lambda v_D}} = I_0 F_2(v_D) \tag{3.70}$$

Now we apply these basic equations to the complete mixing circuit in Figure 3.34, to which two separate voltages are applied, namely, v_A and

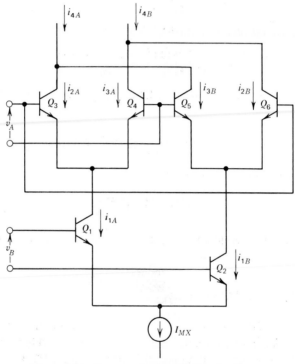

Figure 3.34 Four-quadrant multiplier circuit.

v_B. By direct comparison of the currents in this figure and those in the basic transconductance amplifier of Figure 3.5, we can write the following set of equations:

$$\left.\begin{aligned} i_{1A} &= \frac{I_{MX} e^{\Lambda v_B}}{1 + e^{\Lambda v_B}} = I_{MX} F_1(v_B) \\[2mm] i_{1B} &= \frac{I_{MX}}{1 + e^{\Lambda v_B}} = I_{MX} F_2(v_B) \end{aligned}\right\} \tag{3.71}$$

$$i_{3A} = i_{1A}F_2(v_A) \Big\}$$
$$i_{3B} = i_{1B}F_2(v_A) \Big\}$$
(3.72)

$$i_{2A} = i_{1A}F_1(v_A) \Big\}$$
$$i_{2B} = i_{1B}F_1(v_A) \Big\}$$
(3.73)

$$i_{4A} = i_{2A} + i_{3B} = i_{1A}F_1(v_A) + i_{1B}F_2(v_A) \Big\}$$
$$i_{4A} = I_{MX}[F_1(v_B)F_1(v_A) + F_2(v_B)F_2(v_A)] \Big\}$$
(3.74)

$$i_{4B} = i_{3A} + i_{2B} = i_{1A}F_2(v_A) + i_{1B}F_1(v_A) \Big\}$$
$$i_{4B} = I_{MX}[F_1(v_B)F_2(v_A) + F_2(v_B)F_1(v_A)] \Big\}$$
(3.75)

Now by subtracting Equation 3.75 from 3.74, we have

$$\begin{aligned}
i_{4A} - i_{4B} &= I_{MX}[F_1(v_B)F_1(v_A) + F_2(v_B)F_2(v_A) \\
&\quad - F_1(v_B)F_2(v_A) - F_2(v_B)F_1(v_A)] \\
i_{4A} - i_{4B} &= I_{MX}\left[\frac{e^{\Lambda v_B}e^{\Lambda v_A} + 1 - e^{\Lambda v_B} - e^{\Lambda v_A}}{(1 + e^{\Lambda v_B})(1 + e^{\Lambda v_A})}\right] \\
i_{4A} - i_{4B} &= I_{MX}\left[\frac{e^{\Lambda v_B/2} - e^{-\Lambda v_B/2}}{e^{\Lambda v_B/2} + e^{-\Lambda v_B/2}}\right]\left[\frac{e^{\Lambda v_A/2} - e^{-\Lambda v_A/2}}{e^{\Lambda v_A/2} + e^{-\Lambda v_A/2}}\right] \\
i_{4A} - i_{4B} &= I_{MX}\tanh\frac{\Lambda v_B}{2}\tanh\frac{\Lambda v_A}{2}
\end{aligned}$$
(3.76)

Now recall that the series expansion for $\tanh x$ is

$$\tanh x = x - \frac{x^3}{3} + \frac{2x^5}{17} - \cdots$$
(3.77)

For small v_A and v_B, we will retain only the first term of the expansion, which allows Equation 3.76 to be approximated as

$$i_D = i_{4A} - i_{4B} \approx I_{MX}\frac{\Lambda^2}{4}v_A v_B$$
(3.78)

Thus the difference between these two i_4 currents is directly proportional to the product of the voltages v_A and v_B, and we can see that the multiplication operation is performed. Note that the differential output current i_D has the correct sign for all combinations of input signs. Hence the circuit is called a "four-quadrant" multiplier.

When two sinusoidal-input signals of small amplitude are applied to the multiplier, the differential output current is

$$i_D(t) = \frac{\Lambda^2}{4} I_{MX} V_1 \sin(\omega_1 t) V_2 \sin(\omega_2 t)$$

$$i_D(t) = \left(\frac{\Lambda^2}{8}\right) I_{MX} V_1 V_2 [\cos(\omega_1 - \omega_2)t - \cos(\omega_1 + \omega_2)t]$$

(3.79)

Equation 3.79 shows the familiar heterodyne phenomenon.

The mixer circuit becomes a phase detector if the two input signals have the same frequency and have a phase difference of ϕ. For this case we have

$$i_D(t) = \frac{\Lambda^2}{4} I_{MX} V_1 \sin(\omega t) V_2 \sin(\omega t + \phi)$$

$$i_D(t) = \frac{\Lambda^2}{8} I_{MX} V_1 V_2 [\cos\phi - \cos(2\omega t + \phi)]$$

(3.80)

The frequency-dependent term can be removed with a low-pass filter, leaving the remaining dc component ($\cos\phi$) as a measure of the phase shift existing between the two input signals.

Now consider the special case of the two input signals in quadrature, where

$$\phi = \phi_0 + \Delta\phi \qquad (3.81)$$

The ϕ_0 represents an inherent 90° phase shift deliberately introduced between the two inputs, and $\Delta\phi$ represents any phase departure from this 90° reference value. The dc component of Equation 3.80 can now be written as

$$I_D = K \cos(\phi_0 + \Delta\phi) = -K \sin\Delta\phi \qquad (3.82)$$

where K is a dimensional constant. Since the sine function is odd, changes in $\Delta\phi$ have an associated sign indicating the direction of the phase deviation from the 90° reference phase. If $\Delta\phi$ is restricted to small phase shifts between

$$-\frac{\pi}{6} \leqslant \Delta\phi \leqslant \frac{\pi}{6}$$

we have

$$I_D \approx -\Delta\phi \qquad (3.83)$$

to within 5%. Thus the analog multiplier also serves as a phase detector when an inherent 90° phase shift exists between the two input signals.

3.14 CONTROLLED OSCILLATORS

The function of a controlled oscillator (CO) is to produce an output signal whose frequency can be modulated by either an input voltage or a current signal. The amplitude and wave shape of the CO output are usually of secondary importance to the frequency characteristics of the output signal. In most analog IC circuits, however, the CO output is a square wave having an amplitude suitable for interfacing directly with digital logic circuits. Also in analog ICs, current-controlled oscillators (CCO) are preferred over voltage-controlled oscillators (VCO) for reasons that will become apparent in the subsequent discussion.

One popular type of CCO in analog circuits is the emitter-coupled astable multivibrator, depicted in Figure 3.35. The circuit is essentially a

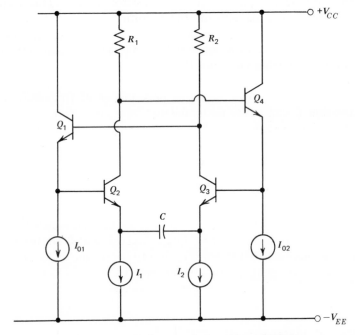

Figure 3.35 Emitter-coupled astable multivibrator.

pair of wide-band, ac-coupled amplifiers in a positive-feedback configuration. The individual amplifiers consist of R_1 and Q_2 for the first and R_2 and Q_3 for the second. The base input terminal of each amplifier is connected to the output of the other (collectors of Q_2 and Q_3) through the emitter followers Q_1 and Q_4, which furnish dc bias and drive capability. The feedback loop is then closed through the capacitor C, establishing a positive-feedback condition that will cause the circuit to oscillate if the loop gain exceeds unity. The frequency of oscillation is inversely proportional to the size of the timing capacitor C and to the current levels in the current sources I_1 and I_2. The size of the capacitor utilized in the circuit sets the "free-running" frequency f_o according to

$$f_o = K_1 \left(\frac{I_1 + I_2}{2} \right) C \qquad (3.84)$$

where K_1 is a dimensional constant. Usually we set $I_1 = I_2$ so that

$$f_o = K_1 \frac{I_1}{C} \qquad (3.85)$$

The constant in this equation contains terms that express the frequency dependence on device parameters and resistor values. This term is generally temperature dependent; the resulting potential temperature dependence of the frequency is usually counteracted by having I_1 vary in a reciprocal way, so that the product of K_1 and I_1 is temperature invariant. The use of a high-quality capacitor then makes the frequency stable with respect to temperature.

Equation 3.85 also shows the linear dependence of frequency on I_1, illustrating that f_o can easily be modulated by a control current.

The output frequency is modulated by applying an input signal to change the magnitudes of I_1 and I_2. For the circuit of Figure 3.35, the output frequency is related to the input current signal by

$$f_{out} = f_o + K_0 i(t) \qquad (3.86)$$

where f_o is the free-running frequency, K_0 is a proportional constant, and $i(t)$ is the input current signal that modulates the individual amplifier currents of I_1 and I_2 in the astable circuit. Since I_1 and I_2 set the astable frequency, an input-current signal is preferred for frequency control over a voltage signal. An input control voltage would have to be converted to a current signal before it could be used to modulate I_1.

3.15 ANALOG SWITCHES

Forming another special class of circuit structures found in analog ICs are the analog switches. These circuits simply act as a short-circuit or an open-circuit path for transmitting or blocking an analog signal between two circuit positions. The shorted or open switch position is usually digitally controlled through appropriate address buffers and decoding schemes.

FETs are used almost entirely as the switching elements in analog switches for analog IC applications. Because of their intrinsic characteristics, FETs enjoy two important advantages over bipolar devices in switching applications. From their physical geometry, FETs are usually fabricated as symmetrical devices, which means that they operate equally well when their source and drain terminals are interchanged. Thus the unipolar FET device has a bipolar signal capability, whereas the ordinary bipolar transistor is capable of only unipolar operation. FETs are therefore equally functional for switching positive and negative signals.

The second principal advantage of FETs is the lack of a dc offset voltage. The practical FET switch has only a series resistance between the input and output without any associated junction potentials that act as dc offsets. When the FET is "on," the series resistance is low (typically 500 Ω); the "off" FET presents a large resistance of typically 50 MΩ.

Of the two types of FETs, the MOS FET is generally preferred over the JFET for switching applications. MOS FETs can be fabricated in the standard analog process discussed in Chapter 1 without requiring the additional diffusion steps necessary for the JFETs. Furthermore, the MOS FETs insulated gate prevents the control voltage used to change the device state from entering the signal path. Finally, much larger reverse-breakdown voltages can be routinely achieved with MOS FETs than with standard-process JFETs.

For switching applications, the enhancement-mode MOS FET is preferred over depletion-mode operation because enhancement devices are self-isolating and can be fabricated by a single diffusion step to form the source and drain regions. All active regions of the MOS FET are reverse biased with respect to one another and to the substrate. Therefore, adjacent devices fabricated on the same substrate are electrically isolated without the need for separate DI or JI techniques.

A single P-channel or N-channel enhancement-mode MOS FET (see e.g., Figure 3.36) can serve as an analog switch. The output characteristics indicate that each device requires a threshold voltage of approximately 3 V to allow current conduction between the source and drain terminals. (The P-channel device requires a negative gate-to-source threshold voltage, whereas a positive voltage is necessary for the N-channel device.) The single MOS FET device best acts as an analog switch when passing unipolar signals. (P-channel MOS FETs are better suited for switching positive signals so that $|V_{GS}|$ always exceeds the threshold voltage. For the same reason, N-channel MOS FETs are better suited for switching negative signals.)

Bipolar analog switching is best accomplished using the complementary switching circuit diagrammed in Figure 3.37. This circuit can pass either

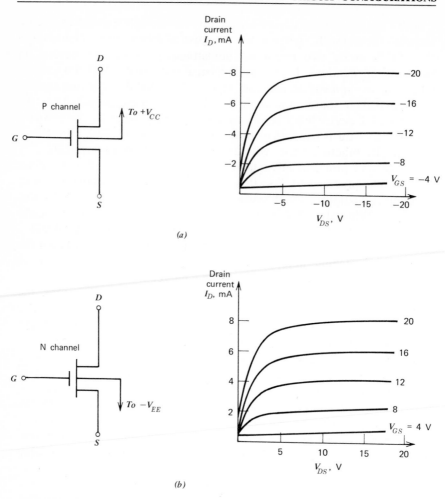

Figure 3.36 Enhancement mode MOS FETs. (a) P channel. (b) N channel.

polarity signal because of the parallel arrangement. The switch is "shorted" by applying a positive control voltage greater than the threshold voltage to the gate of the N device and a negative voltage greater than V_T to the gate of the P device. Large positive analog signals will pass through the P device and negative signals through the N device. The drain resistance of each device is a function of the polarity and amplitude of the input signal. However, because of the complementary action of the circuit, when the channel resistance of one device decreases, the resistance of the other device increases. Thus to a first-order

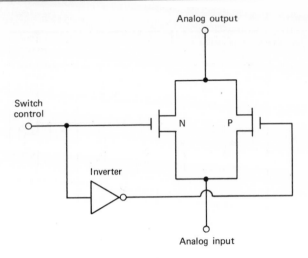

Figure 3.37 Complementary MOS FET switching circuit.

approximation, the effective resistance of the parallel combination remains relatively constant for bipolar input signals of any amplitude.

The complementary circuit does suffer from one serious shortcoming. Input voltages that exceed either power-supply limit will turn "on" the source–body diode of the respective MOS FET. This will result in a low-resistance path between the input and one of the power supplies, which could destroy the device under excessive overvoltage conditions.

Figure 3.38 is a circuit configuration that will protect the complementary MOS (CMOS) switching devices Q_1 and Q_2 from input overvoltage. The circuit will prevent input signals from exceeding either supply voltage by clamping the input of Q_1 and Q_2 at $(-V_{EE}-V_{BE,Q5})$ and $(+V_{CC}+V_{BE,Q6})$, respectively. The protection circuit is symmetrical; therefore, we can explain the operation for positive overvoltages knowing that a similar action occurs for the negative case.

The positive-overvoltage circuit consists of Q_6, D_6, D_7, and Q_7. The operation of this circuit depends on the source–body diodes of Q_2, which are normally reverse biased with small signal levels. These diodes are shown in Figure 3.38b. Let us first consider the case of an "off" switch (Q_1 and Q_2 both "off"). Under this condition, Q_7 will be fully "on." When the input voltage exceeds $+V_{CC}$, the protection circuit must activate before the source–body diode of Q_2 conducts. That is, if this diode is permitted to conduct, transistor action to the output will effectively turn "on" the Q_2 switch. Since the base–emitter junction of Q_6 and the

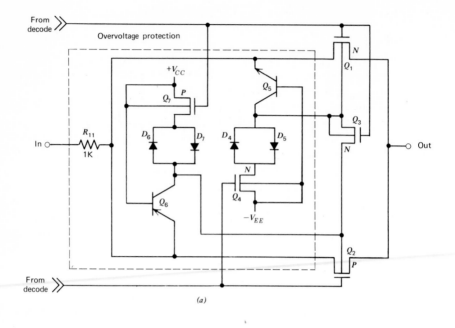

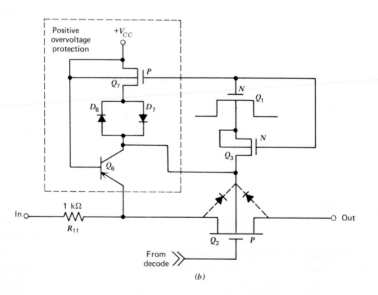

Figure 3.38 CMOS analog switch. (*a*) Complete circuit. (*b*) Positive overvoltage protection circuit.

source-body diode have near-equal forward voltage potentials (~0.6 V), it might appear at first glance that a race condition exists between Q_6 to turn "on" and the source–body diode of Q_2 to also turn "on." A closer look reveals that diode D_6 is in series with the source–body diode and offers an additional diode drop. Therefore, to turn "on" the source–body diode, the input voltage must be at least two diode potentials greater than $+V_{CC}$ to turn "on" Q_6, preventing the occurrence of a race condition. Diode D_7 provides collector bias to Q_6, causing a $V_{CE(sat)}$ of approximately 0.1 V to exist during the overvoltage condition. This voltage across the source–body diode assures that an "off" condition is maintained over the specified analog overvoltage range. The series 1 kΩ resistor limits the current flowing through the protection circuit under extreme overvoltage conditions.

Overvoltage protection is also realized in the case of an "on" switch. In this case Q_3, as well as Q_1 and Q_2, will be "on" while Q_7 and Q_4 will be "off." Transistors Q_7 and Q_4 are turned "off" to allow both N- and P-channel bodies to float with the analog input voltages. This scheme eliminates source–body modulation and results in a constant "on" resistance value over the specified analog range. Positive and negative overvoltages are clamped at one diode junction above and below the $+V_{CC}$ and $-V_{EE}$ supply levels, respectively. The analog switch we have just described forms the basic circuit for multiplexer applications treated in Chapter 7.

3.16 RESISTOR NETWORK CONFIGURATIONS

Specialized resistor networks intended for switching applications are used primarily in digital-to-analog (D/A) and analog-to-digital (A/D) converters. These resistor networks generally are of two types: binary weighted, and R–$2R$ ladder configurations. In this section we describe the special characteristics of each of these networks.

Binary-Weighted Resistor Networks

Probably the most straightforward ladder network is the binary-weighted resistor type, presented in Figure 3.39. Each resistor in this circuit is twice as large as the preceding resistor and one-half as large as the resistor that follows. In general, then, for any one resistor we have

$$R_i = 2^{i-1}R_1, \qquad i = 1, 2, 3, 4, \ldots, n \qquad (3.87)$$

The smallest resistor in the ladder network is R_1 and the largest $2^{n-1}R_1$ where n is the total number of resistors in the network. The switches in

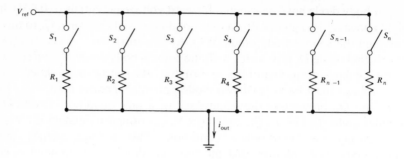

Figure 3.39 Binary-weighted resistor network.

Figure 3.39 represent FET devices of the type described in Section 3.15. It should be obvious that by opening and shorting the various switches, we can generate an output current that is a binary quantization of the applied reference voltage V_{ref}.

Binary-weighted networks are not well suited for monolithic circuit implementation for several reasons. If more than about four or five resistors are needed in the ladder network, the size (both physical and electrical) of the largest resistor quickly becomes impractical unless the chip area is reduced by a specialized technique (e.g., a pinched resistor). However, the low accuracy of pinched resistors precludes their use in ladder networks. Another shortcoming of binary-weighted networks is that all resistors must have tight tolerances. For example, if n binary-weighted resistors are used in a network, the tolerance for the largest valued resistor must exceed $100\%/2^n$, and it is difficult to achieve this high tolerance in monolithic circuits for $n \geqslant 4$.

R–$2R$ Resistor Ladder Networks

The R–$2R$ network is well suited to monolithic circuits because the configuration requires only two different resistor sizes (i.e., R and $2R$). Furthermore, as we will show, the output impedance of this ladder network is constant regardless of the number of stages used.

The R–$2R$ ladder network appears in Figure 3.40. This network employs twice as many components as the binary-weighted network. However, since all components have a value of either R or $2R$, excessively large resistor values are not necessary. Furthermore, the close matching among resistors that is possible with monolithic structures improves the accuracy of the network.

We now show that the R–$2R$ network produces an output voltage having a binary relationship to the reference voltage V_{ref}. To evaluate

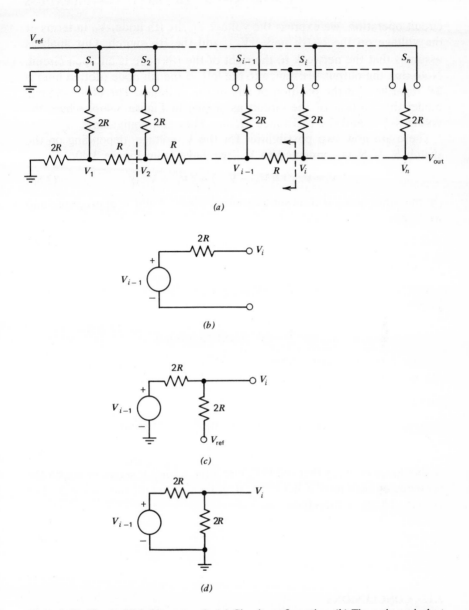

Figure 3.40 The $R-2R$ ladder network. (a) Circuit configuration. (b) Thevenin equivalent at the ith node. (c) Thevenin equivalent with $S_i = 1$. (d) Thevenin equivalent with $S_i = 0$.

circuit operation, we express the voltage on the ith node, V_i, in terms of the voltage on the $i-1$th node, V_{i-1}, and the ith switch. Our analysis assumes that the network to the right of the ith node is an open circuit. Note that the output resistance to the left of the ith node (dotted line) is $2R$ regardless of the position of any of the switches. Therefore, we can model this portion of the circuit as shown in Figure 3.40b, where the voltages V_{i-1} and the resistance $2R$ are Thevenin equivalents.

There are now two possibilities for the V_i voltage, depending on the switch S_i. If S_i is connected to V_{ref}, we have from Figure 3.40c

$$V_i = V_{i-1} + \tfrac{1}{2}(V_{\text{ref}} - V_{i-1}) = \tfrac{1}{2}(V_{\text{ref}} + V_{i-1}) \tag{3.88}$$

On the other hand, if S_i is set to ground, Figure 3.40d is appropriate and we have

$$V_i = \tfrac{1}{2}V_{i-1} \tag{3.89}$$

Therefore, we can conclude that

$$V_i = \tfrac{1}{2}(V_{i-1} + S_i V_{\text{ref}}) \tag{3.90}$$

where

$$S_i = \begin{cases} 1, & S_i \text{ set to } V_{\text{ref}} \\ 0, & S_i \text{ set to ground} \end{cases}$$

Now if we move to the output node where $V_i = V_n = V_{\text{out}}$, the next switch S_{n+1} is obviously open because it is nonexistent. Therefore, we can write

$$V_{\text{out}} = V_i = \tfrac{1}{2}(S_i V_{\text{ref}} + V_{i-1}) \tag{3.91}$$

But since we can infer V_{i-1} from Equation 3.90, we now put

$$V_{\text{out}} = \tfrac{1}{2}(S_i V_{\text{ref}}) + \tfrac{1}{4}V_{i-2} \tag{3.92}$$

It should be obvious that we are generating a binary series in which the presence of each term is determined by the position of the switch S_i. The output voltage is therefore

$$V_{\text{out}} = V_{\text{ref}} \sum_{i=1}^{n} \frac{S_i}{2^{n-i+1}} \tag{3.93}$$

3.17 CONCLUSIONS

The analog circuit configurations presented and discussed in this chapter lend themselves well to monolithic integration. Basic circuit structures that have become building blocks to the design engineer were illustrated in the first eight sections. Among those described were differential amplifiers, constant-current circuits, biasing networks, emitter followers, and basic gain stages.

Certain improvements in performance are possible through modifications to the basic circuits. These modifications lead to the specialized circuits, which were presented in the remaining sections. Among the circuits described were admittance cancellation techniques, circuits utilizing FETs, analog mixing circuits, controlled oscillators, analog switches, and resistor networks. As we study complete, working devices used in specific applications in the chapters that follow, we will have many occasions to refer back to these basic and specialized circuit structures.

BIBLIOGRAPHY

1. Ahmed, H., and P. J. Spreadbury, *Electronics for Engineers, An Introduction*, Cambridge University Press, London, 1973.

2. Angelo, E. James, Jr., *Electronics: BJTs, FETs, and Microcircuits*, McGraw-Hill, New York, 1969.

3. Belove, Charles, Harry Schachter, and Donald L. Schilling, *Digital and Analog Systems, Circuits, and Devices: An Introduction*, McGraw-Hill, New York, 1973.

4. Chirlian, Paul M., *Integrated and Active Network Analysis and Synthesis*, Prentice-Hall, Englewood Cliffs, N.J., 1967.

5. Deboo, Gordon J., and Clifford N. Burros, *Integrated Circuits and Semiconductor Devices: Theory and Application*, McGraw-Hill, New York, 1971.

6. Eimbinder, Jerry, *Semiconductor Memories*, Wiley, New York, 1971.

7. Fitchen, Franklin C., *Electronic Integrated Circuits and Systems*, Van Nostrand-Reinhold, New York, 1970.

8. Graeme, Jerald G., Gene E. Tobey, and Lawrence P. Huelsman, *Operational Amplifiers: Design and Applications*, McGraw-Hill, New York, 1971.

9. Gray, Paul, E., and Campell L. Searle, *Electronic Principles: Physics, Models, and Circuits*, Wiley, New York, 1969.

10. Grebene, Alan B., *Analog Integrated Circuit Design*, Van Nostrand-Reinhold, New York, 1972.

11. Hunter, Lloyd P., *Handbook of Semiconductor Electronics*, McGraw-Hill, New York, 1970.

12. Lathi, B. P., *Random Signals and Communication Theory*, International Textbook, Scranton, Pa., 1968.

13. Le Page, W. R., and S. Seely, *General Network Analysis*, McGraw-Hill, New York, 1952.

14. Meyer, Charles S., David K. Lynn, and Douglas J. Hamilton, *Analysis and Design of Integrated Circuits*, McGraw-Hill, New York, 1968.

15. Millman, Jacob, and Christos C. Halkias, *Electronics Devices and Circuits*, McGraw-Hill, New York, 1967.

16. Millman, J., and C. C. Halkias, *Integrated Electronics: Analog and Digital Circuits and Systems*, McGraw-Hill, New York, 1972.

17. Millman, J., and Herbert Taub, *Pulse, Digital, and Switching Waveforms*, McGraw-Hill, New York, 1965.

18. Motchenbaucher, C. D., and F. C. Fitchen, *Low Noise Electronic Design*, Wiley, New York, 1973.

19. Pierce, J. F., and T. J. Paulus, *Applied Electronics*, Merrill, Columbus, Ohio, 1972.

20. Pierce, J. F., *Semiconductor Junction Devices*, Merrill, Columbus, Ohio, 1967.

21. Pierce, J. F., *Transistor Circuit Theory and Design*, Merrill, Columbus, Ohio, 1963.

22. RCA Inc., "Linear Integrated Circuits," RCA Corp., Technical Series IC-42, 1970.

23. RCA Inc., "Linear Integrated Circuits and MOS Devices," RCA Corp. No. SSD-202, 1972.

24. Ryder, John D., *Electronic Fundamentals and Applications*, Prentice-Hall, Englewood Cliffs, N.J., 1970.

25. Schilling, Donald L., and Charles Belove, *Electronic Circuits: Discrete and Integrated*, McGraw-Hill, New York, 1968.

26. Schwartz, Seymour, *Integrated Circuit Technology*, McGraw-Hill, New York, 1967.

27. Solomon, J. E., W. R. Davis, and P. L. Lee, "A Self-Compensated Monolithic Operational Amplifier with Low Input Current and High Slew Rate," IEEE International Solid State Circuits Conference, Philadelphia, February 19, 1969, pp. 14–15.

28. Stewart, Harry E., *Engineering Electronics*, Allyn & Bacon, Boston, 1969.

29. Warner, Raymond M., Jr., and James N. Fordemwalt, *Integrated Circuits: Design Principles and Fabrication*, McGraw-Hill, New York, 1965.

CHAPTER **4**

VOLTAGE COMPARATORS

In the preceding chapter we learned how individual analog circuits are designed to accomplish specific circuit functions. Now we can begin to interconnect several of these basic building-block configurations, to produce a device having terminal characteristics useful for a variety of general applications. The first device we examine is the voltage comparator.

4.1 FUNCTIONAL OPERATION

A voltage comparator (VC) senses the relative polarity of the differential voltage applied to the comparator's two inputs. The output of an ideal VC will be at a voltage level corresponding to a logic 1 whenever the difference voltage existing between the noninverting and inverting inputs is positive, or

$$V_{out} = \text{logic 1} \qquad \text{when} \quad (V_+ - V_-) > 0$$

When this difference voltage is negative, the VC output is a logic 0, or

$$V_{out} = \text{logic 0} \qquad \text{when} \quad (V_+ - V_-) < 0$$

The output of an ideal comparator changes state when $V_+ = V_-$ and has the input-output characteristic shown in Figure 4.1. In its simplest form, a single comparator could be considered to be a one-bit, analog-to-digital converter, since the VC output represents a two-state digital quantization of the relative polarity of the analog input. As such, comparators frequently serve to interface between analog and digital systems. Analog transducers of many types, such as photodiodes, magnetometers, and radiation detectors, produce low-level signals that are converted into

119

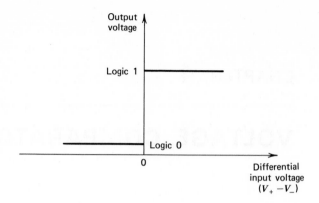

Figure 4.1   Input–output characteristics of an ideal voltage comparator.

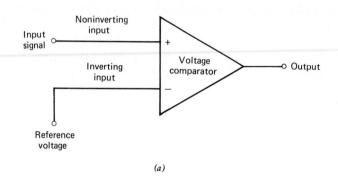

(a)

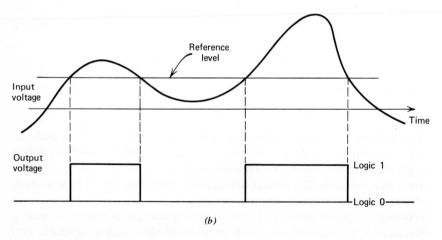

(b)

Figure 4.2   The voltage comparator. (a) Symbolic representation. (b) Voltage wave shapes.

120

digital information. Comparators detect these low-level signals and convert them into binary logic levels.

An important application of a VC is level detection. A reference voltage and an input signal are separately applied to the VC inputs, as in Figure 4.2. (Power-supply connections are not shown but are presumed to be present.) The VC output is at a low voltage level (logic 0) whenever the input is more negative than the reference voltage; it is at a high level (logic 1) whenever the input is more positive than the reference.

The VC is equally functional if the input- and reference-voltage terminals are interchanged. This reversal of terminals will simply invert the output-voltage wave shape.

With the VC there is no intention of reproducing any part of the original input-signal wave shape. Therefore most VCs are essentially high-gain differential amplifiers operating in the open-loop manner and driving an output stage conditioned for specific logical voltage levels. As we shall see later, operational amplifiers can serve as comparators by clipping their output voltage to the required logic levels. However, because op amps are designed to preserve a linear relationship between input and output signals, they frequently have response times in the tens of microseconds, which may be too slow for many comparator applications. We now investigate the arrangement of the various building-block circuits used to produce an IC voltage comparator.

4.2 VC CIRCUIT ORGANIZATION

Voltage comparators generally are structured using three of the basic circuit configurations we dealt with in Chapter 3. A typical interconnection scheme appears in Figure 4.3. The input stage of the VC is a high-gain, direct-coupled difference amplifier. Like most analog IC devices, the terminal characteristics of this differential input amplifier determine the practical limitations to ideal VC operation. Ideally the VC would present no loading effect on the attached circuit. This is not possible practically because the input stage is almost always direct coupled, and some small dc current must exist between the signal sources and the positive and negative inputs of the VC to establish proper bias of the input transistors. Failure to provide a dc current path for biasing the input stage is a frequent oversight in initial designs, and it prevents proper operation of the internal biasing network.

Some external control over the difference amplifier is provided by the offset adjustment. These control terminals allow compensation for slight device mismatches which typically exist in the transistors in the input

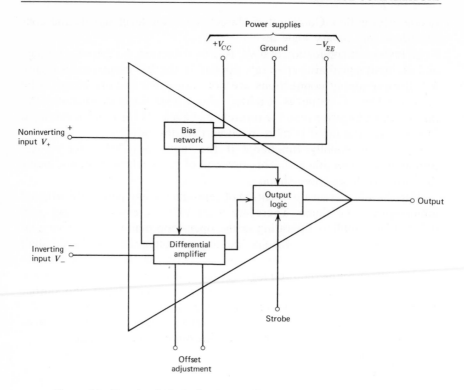

Figure 4.3 Functional block-diagram organization of a voltage comparator.

stage. These device mismatches cause a small offset voltage V_{IO} to be generated between the positive and negative terminals of a practical comparator. The transition voltage at which the comparator changes logic state becomes shifted from zero to V_{IO} in accordance with

$$V_{out} = \text{logic } 1 \quad \text{when} \quad (V_+ - V_-) > V_{IO}$$

and

$$V_{out} = \text{logic } 0 \quad \text{when} \quad (V_+ - V_-) < V_{IO}$$

Connection of an external network containing a variable resistance between the power-supply and the offset-adjust terminals will permit the adjustment of the point at which the transition between logic levels begins within typically 1 mV of zero. We examine these practical limitations further in the next section.

The biasing network (attached to the difference amplifier in Figure 4.3) is a highly stable, temperature-compensated network designed to achieve optimum current levels in the various transistors over a wide range of input voltages. The bias network also establishes the proper currents and voltage levels in the output logic stage. The function of this stage is to produce logic voltage levels compatible with a particular logic type (e.g., transistor-transistors logic—TTL, and emitter-coupled logic—ECL) with which the VC is intended to interface. The same output characteristics are used to define the operation of this logic stage as those associated with the various digital logic types. Detailed treatments of these characteristics may be found in the literature (5, 9, 12, 14, 16, 17, 24).

We now briefly examine the circuit diagram of a versatile voltage comparator to show typical connections and interactions of these circuit structures. The circuit is the comparator from the HA-2111 series (see Figure 4.4). If we take a general overview of the complete circuit, several interesting features stand out. For example, a strobing input allows the circuit to be disabled through the application of an external logic signal. A logic 1 applied through an external, inverting transistor whose collector is attached to the strobe terminal will disable the comparator, making the output high regardless of the polarity of the difference signal input. This feature can be extremely useful for comparator applications in which initial transients appear on the input signal. Unless the comparator is disabled until these transient effects die out, the VC output may change state several times, making it impossible to distinguish between the true input signal level and the initial transients.

Another interesting feature utilized in this voltage comparator is the PNP transistors Q_1 and Q_2, which drive the standard NPN difference amplifier stage composed of Q_3 and Q_4. These PNP transistors act as buffers to the differential stage to reduce the bias current of the NPN stage by approximately a factor of β. The switching speed is not seriously degraded by the presence of these PNP devices because of the emitter-follower configuration chosen. This configuration also permits a wide range of reference voltages to be applied to the inputs, approaching, and only limited by, the positive and negative power supplies.

The differential output from the Q_3–Q_4 stage is further amplified by the Q_8–Q_9 differential pair. (See the simplified circuit of Figure 4.4b.) Current-source biasing is used throughout to set the quiescent current and to allow for a wide variety of supply voltages without affecting circuit performance.

The output stage is designed to be very flexible, to provide voltage levels compatible with RTL, DTL, TTL, and MOS logic types. Appropriate logic levels can be selected through external connections to the open-collector output transistor Q_{15}. The open-collector output can be returned to the positive supply through a pullup resistor, or it can be used as a point to switch loads that are connected to a voltage higher than the positive supply voltage. This open-collector configuration enables direct connection of relays and lamp loads. Q_{16} and R_{13} serve for short-circuit protection of the output transistors (from excessive load currents). The comparator will operate from a single positive supply voltage if the negative supply is grounded. However, using a negative supply will increase the common-mode range of the circuit.

The voltage comparator can be converted to a buffer amplifier through another arrangement of external connections to the output stage. Buffer amplifiers, also commonly called isolation amplifiers or emitter followers, are used to prevent loading of the signal source while increasing the signal drive capability. Referring to Figure 4.4b, the "output" is connected directly to the positive voltage power supply, and the output is taken

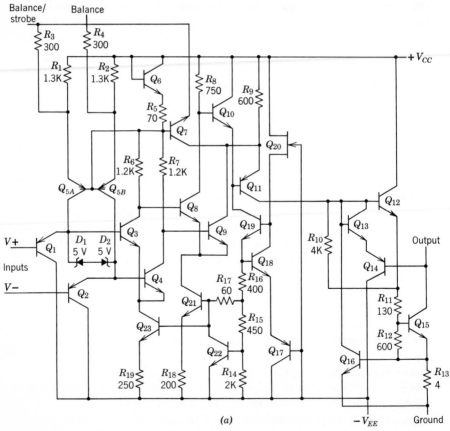

Figure 4.4 The HA-2111 voltage comparator. (*a*) Complete circuit. (*b*) Simplified schematic.

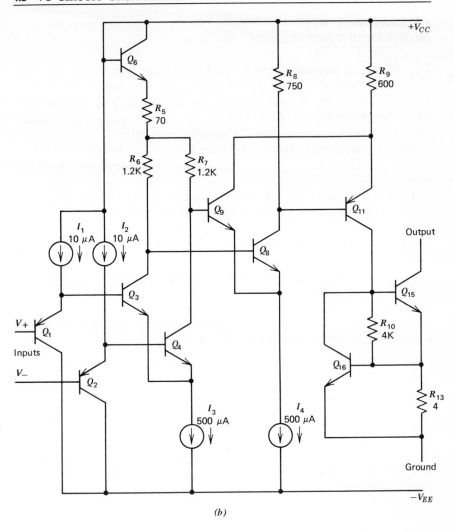

(b)

across a load resistor inserted between the external VC terminal denoted as "ground" and the true system ground. The output transistor Q_{15} provides no signal inversion for this emitter-follower connection because the output is taken at the emitter instead of at the collector. Therefore, there is a polarity reversal between the positive and negative input terminals (i.e., the positive input becomes the negative input, and vice versa, for the emitter-follower configuration).

4.3 PRACTICAL DESIGN CONSIDERATIONS OF VOLTAGE COMPARATORS

The effects of nonideal operation of a voltage comparator are best reflected in the performance indexes and parameters cited on the device's specification sheet. In this section we discuss several of the more important parameters to internal device and circuit operation and indicate the effects of these parameters in specific applications.

Since we are already familiar with the HA-2111 comparator, let us use this device's specifications as a typical example of comparator perform-ance.

Table 4.1 summarizes the performance of the HA-2111 comparator when using a single 5 V supply. The comparator will operate with supply voltages up to ±15 V with a corresponding increase in the range of acceptable input voltages. Other characteristics in the table are essentially unchanged at the higher power-supply voltages.

The performance indexes of Table 4.1 are of two principal types: dc and ac. Indexes reflecting on the static operation are considered to be dc; those which describe the device under switching conditions are consid-ered to be ac. All the important parameters cited in Table 4.1 are dc parameters except response time which is an ac index.

Table 4.1 Important electrical characteristics of the HA-2111 voltage comparator when operating from a single 5 V supply ($T_A = 25°C$)

	Limits			
Parameter	Minimum	Typical	Maximum	Units
Input-offset voltage[a]		0.7	3	mV
Input-offset current		4	10	nA
Input-bias current		60	100	nA
Voltage gain		200		V/mV
Response time[b]		200		ns
Common-mode range	0.3		3.8	V
Output-voltage swing			50	V
Output current			50	mA
Fanout (DTL/TTL)	8			
Supply current		3	5	mA

[a] The offset voltages and offset currents given are the maximum values required to drive the output within one volt of either supply with a 1 mA load. Thus these parameters define an error band and take into account the worst-case effects of voltage gain and input impedance.
[b] 100 mV input step; 5 mV overdrive.

As we have pointed out previously, the input-offset voltage is a measure of the magnitude of the actual difference voltage existing between positive and negative inputs which causes the output voltage to indicate a state change. Since the HA-2111 has a flexible output stage designed to interface with several logic types, a definition of switching occurrence is necessary. The definition applied is one which considers that the output has changed state when the output voltage is within 1 V of the logic and ground supply levels utilized in the output stage when sourcing or sinking 1 mA of load current (see Figure 4.4). Figure 4.5

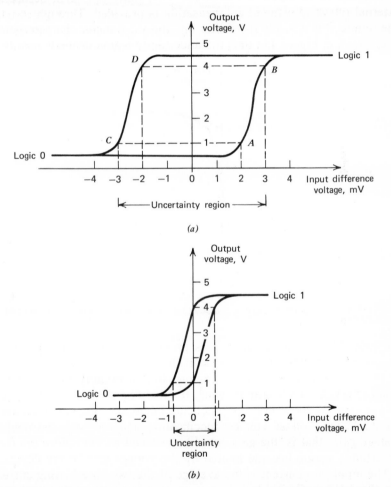

(a)

(b)

Figure 4.5 Effects of a nonzero input-offset voltage on the switching characteristics of a comparator for (a) 3 mV offset voltage, (b) 0.7 mV offset voltage.

illustrates these switching characteristics for the maximum (3 mV) and typical (0.7 mV) offset voltages indicated in Table 4.1. The output supplies are considered to be +5 V and ground; thus state changes occur at +4 V and +1 V. Figure 4.5a shows that the output of the comparator is absolutely at the logic 1 level for a +3 mV differential input voltage (point B) and absolutely at a logic 0 (point C) for a −3 mV differential input voltage, thereby presenting a 6 mV uncertainty region. The uncertainty region is considerably reduced when the typical offset voltage of 0.7 mV is considered, as in Figure 4.5b.

Both curves in Figure 4.5 illustrate comparator operation before any external offset adjustment compensation is provided. Through resistive fine tuning, it is possible to "center up" the VC transfer characteristic to that shown in Figure 4.6; thus the uncertainty region depends mainly on

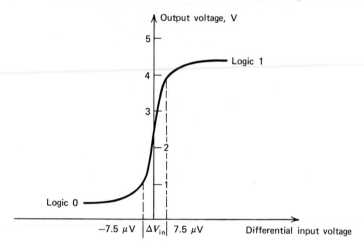

Figure 4.6 Comparator transfer characteristic when properly externally trimmed.

the comparator's voltage gain. This produces the minimum uncertainty region that is practically attainable. If the typical 200 V/mV voltage gain of the HA-2111 is used as an example, the uncertainty range is a very small 15 μV for a 3 V output swing. It should be pointed out that with a gain as large as 200 V/mV, the practical limitation to VC operation lies in the 3 mV input-offset voltage specification and not with the device's voltage gain; that is, the gain could be reduced to 500 before the 6 mV uncertainty region became limited by the voltage gain of the device.

The input-bias current is the average of the two base-biasing currents for the PNP buffering transistors, connected to the differential input stage. The low typical value of 60 nA quoted in Table 4.1 is indicative of

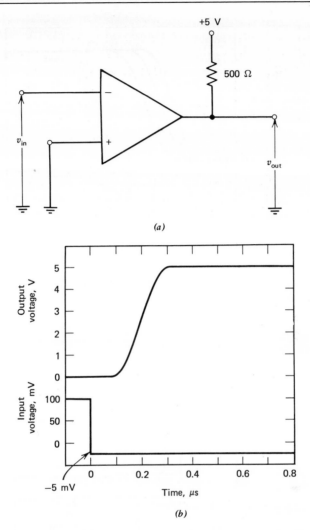

Figure 4.7 Test conditions for determination of response time of voltage comparator. (*a*) Test circuit. (*b*) Voltage waveshapes for 5 mV input overdrive.

the extremely small loading effect this comparator presents to a signal source.

The input-offset current reflects the degree of matching achieved between the two input transistors. Being the absolute value of the difference between these two input transistors' base currents, the offset current is indicative of the close device matching ($\approx 3\%$) achievable in the fabrication processes.

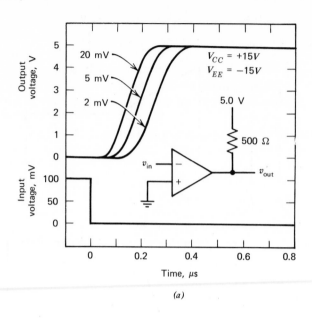

(a)

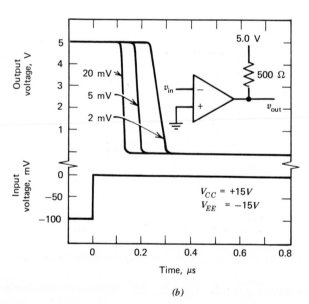

(b)

Figure 4.8 Response and delay time variations for various overdrives. (a) For 0 to 1 output transition. (b) For 1 to 0 output transition.

130

One important index of the dynamic switching characteristics of the VC is the response time. The response time of 200 ns cited in Table 4.1 is not a constant value but varies in accordance with the input-signal test conditions. For example, this 200 ns value is the time necessary for the VC output to change state when a 100 mV input signal with a 5 mV overdrive is applied, as illustrated in Figure 4.7. The output voltage exhibits a delay time of approximately 100 ns after the application of the input signal before beginning the 0 to 1 transition. Once initiated, this state transition requires 200 ns, given as the typical response time.

The response and delay times are influenced strongly by the input-signal test conditions. The 5 mV overdrive is the differential driving signal, since the noninverting input terminal is grounded. The response and delay times varies inversely with input overdrive voltage (see Figure 4.8). (The overdrives of the input-voltage wave shapes are not shown but are understood to be the amplitude excursions measured from ground reference.) Figure 4.8*b* gives the transient behavior in the reverse direction (i.e., for a 1-to-0 output transition).

4.4 APPLICATIONS OF VOLTAGE COMPARATORS

We now illustrate several important applications of voltage comparators to accomplish specific system functions. In addition to the level detector previously considered, we show comparator applications to generate a zero-crossing detector, a window detector, and a Schmitt trigger.

Zero-Crossing Detector

Frequently when processing analog data, a power spectrum analysis of some particular signal wave form must be performed. To accomplish this, the frequency content of the signal wave form of interest must be extracted from the broadband signal and noise within a specified pass-band. A zero-crossing detector using a comparator provides a simple and effective means of performing such an operation. A zero-crossing detector will produce an output signal that changes state each time the analog input signal passes through a reference voltage of zero volts. The input signal is thus inverted into a train of frequency-dependent pulse widths, and the resultant zero-crossing intervals can be examined for frequency content.

Zero sensing of the input signal virtually eliminates distortion caused by amplitude fluctuations, circuit variations, and noise, and permits simplification of further data processing through the use of digital techniques. The zero-crossing detector is an extremely valuable circuit for systems that require information processing, signal storage, and signal correlation.

The circuit in Figure 4.9 is a useful zero-crossing detector having the additional feature of strobe capability through the application of an external logic signal. The connection indicated in this figure produces an

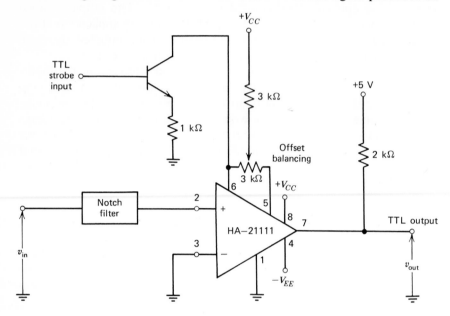

Figure 4.9 Zero-crossing detector.

output that is DTL–TTL compatible. Since the input is applied to the noninverting terminal with the inverting terminal grounded, the output will be in the logic 1 state whenever the input is positive and in the 0 state whenever the input is negative. The output changes state each time the input signal crosses the zero reference level. Some frequency filtering is done via the notch filter, which selectively removes unwanted frequencies and accepts for analysis the signals that are within a specified passband.

A modification of the zero-crossing detector for driving MOS logic circuits is presented in Figure 4.10. In this circuit both a positive power supply and a negative power supply are required for the circuit to work with zero common-mode voltage. The output logic levels for the +5 and −10 V supplies are approximately 4.5 V for the logic 1 state and −9.5 V for the logic 0 state. It should be pointed out that both circuits in Figures 4.9 and 4.10 require both positive and negative supply voltages to be used if true zero sensing is to be attained.

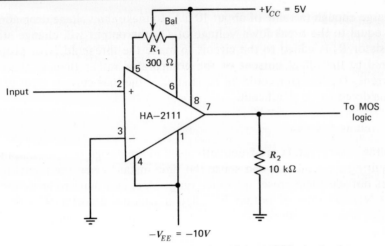

Figure 4.10 Zero-crossing detector for driving MOS logic circuits.

Level Detector

Slight modifications to the circuits of Figures 4.9 and 4.10 permit sensing at levels other than zero volts. Such an example appears in Figure 4.11, in which a precision level detector for a photodiode is acting as the sensing device. Zener diode D_1 sets up a stable reference voltage to be applied to one input of the comparator. When the current through the photodiode D_2

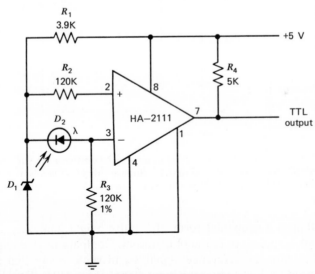

Figure 4.11 Precision level detector for photodiode. (Reprinted courtesy of National Semiconductor Corporation from *Linear Applications Handbook*.)

is large enough (a value of about 10 μA) to make the voltage drop across R_3 equal to the breakdown voltage of D_1, the output will change state. Resistor R_2 is added to the circuit to make the threshold error proportional to the offset current of the comparator rather than to the bias current. This resistor could be eliminated if the bias-current error is not considered to be significant.

Window Detector

Voltage comparators are frequently used in the design of test equipment requiring a circuit that can sense the time instant when the input signal goes outside some preset tolerance range. A circuit that produces such a GO–NO-GO type of output is called a window detector. Figure 4.12 shows how two voltage comparators are employed to produce a window

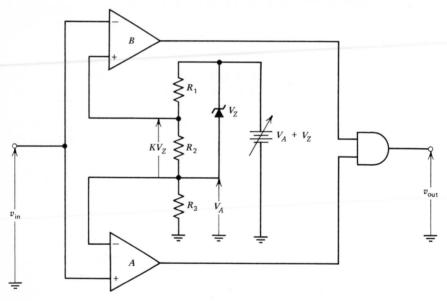

Figure 4.12 Window comparator for detecting input signals between the limits of $V_A \leq v_{in} \leq V_A + KV_Z$, where $K = R_2/(R_1 + R_2)$. (From L. Strauss, *Wave Generation and Shaping*. Copyright 1970 by McGraw-Hill Book Company. Used with permission of McGraw-Hill Book Company.)

detector that is a simplified form of a pulse-height analyzer. The same input signal is applied to both comparators. Comparator A compares the input signal with the reference signal V_A in such a way that a logic 1 output is produced whenever the input signal exceeds V_A. Comparator B, being inversely connected, compares the same input signal with a larger reference voltage, namely, $V_A + KV_Z$. The B comparator produces a logic 1 output only when the input signal is less than $V_A + KV_Z$. The outputs from the two comparators are ANDed, yielding a logic 1 whenever both comparators are at the logic 1 level. However, this occurs only for input voltages in the range of $V_A \leq v_{in} \leq V_A + KV_Z$. The circuit we have examined produces a window that indicates by a logic 1 output the presence of an input signal falling within the range preset by the V_A and KV_Z voltages. Higher input levels can be detected by adding a voltage divider at the input. Also, the circuit could be made to accomplish pulse-height analysis by slowly varying V_A while still maintaining appropriate KV_Z separation.

Schmitt Trigger

The noise levels riding on low-level input signals can be particularly troublesome in level detectors. For example, when the input signal approaches the reference level, noise can cause the comparator to change state incorrectly. If the amplitude of the input signal is exactly equal to the reference voltage, the comparator will change state randomly in response to the polarity of the noise voltage. To avoid this difficulty, positive feedback is frequently employed to alter the comparator's transfer characteristic (see Figure 4.13). The modified circuit, known as a Schmitt trigger, produces two voltages V_2 and V_1, which are the levels at which the comparator changes state: V_2 and V_1 are called the upper and lower trigger points, respectively. The difference between V_2 and V_1 is the hysteresis voltage, which is made somewhat larger than the maximum expected noise voltage. With this hysteresis included, the comparator is less sensitive to noise and responds only to changes in the input-signal level.

The amount of hysteresis obtainable is determined by the ratio of the two resistors R_1 and R_2 and by the full voltage swing of the comparator or $V_{OH} - V_{OL}$. The upper and lower trigger points can be calculated from the following equations:

$$V_2 = V_{ref} \frac{R_1}{R_1 + R_2} + V_{OH} \frac{R_2}{R_1 + R_2} \tag{4.1}$$

$$V_1 = V_{ref} \frac{R_1}{R_1 + R_2} + V_{OL} \frac{R_2}{R_1 + R_2} \tag{4.2}$$

The hysteresis voltage, the difference between the upper and lower trigger points, reduces to

$$V_H = \frac{R_2}{R_1 + R_2} (V_{OH} - V_{OL}) \tag{4.3}$$

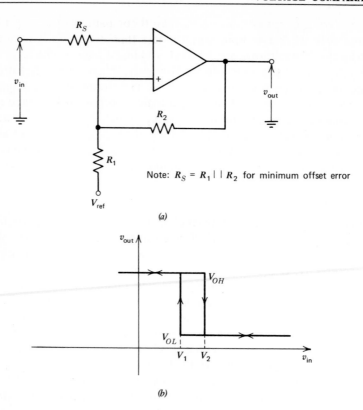

Note: $R_S = R_1 \mid\mid R_2$ for minimum offset error

(a)

(b)

Figure 4.13 Schmitt trigger. (a) Circuit with positive feedback. (b) Transfer characteristic showing hysteresis effect.

The major advantage of this type of comparator is its inherent self-latching action. Since the upper and lower trigger points are separated, the circuit has a built-in noise immunity equal to the width of the hysteresis. Once the comparator changes states, small perturbations will not cause reswitching and consequent erratic indications. The price paid for this broadening of the input transition zone is the inability to narrowly define a single-voltage comparator level.

Analog-Voltage–Digital-Pulse Converter

Figure 4.14 shows the application of a voltage comparator to perform an analog-voltage-to-digital-pulse conversion. The voltage comparator compares the input signal with a linear ramp being generated and shown as voltage v_1. As long as the input voltage is greater than the ramp voltage,

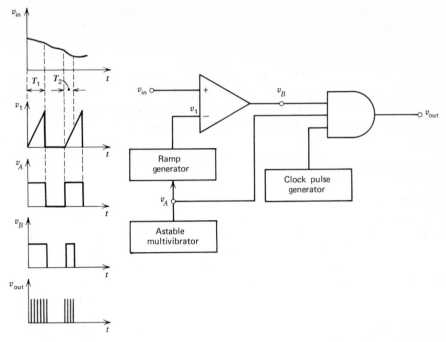

Figure 4.14 Analog-voltage-to-digital-pulse converter. (Taken and modified from L. Strauss, *Wave Generation and Shaping*, p. 460.)

the comparator's output will be at the logic 1 level. At some time T_1, the two signals are equal and the comparator output drops to a logic 0 level. The duration of the logic 1 signal depends on the amplitude of the input signal. The comparator serves to convert the input signal's amplitude into a proportional time interval.

At the time instant when the linear ramp first begins rising, the astable-multivibrator output is applied to one input of the AND gate. The coincidence of the comparator output and the astable-multivibrator's output opens the AND gate and permits transmission of a train of positive clock pulses. The specific number of pulses transmitted through the AND gate is directly proportional to the time the gate remains open. This time, in turn, is proportional to the input voltage amplitude. Thus a discrete number of output pulses are counted, stored, and related to the input amplitude of the applied voltage signal.

The sampling time depends on the multivibrator period. The "on" time must be long enough for the ramp to rise from zero to the largest input voltage expected. The "off" time, which determines the sampling interval, is set by the rate of change of the input voltage. Rapidly varying input signals require a higher sampling rate. When the sampling time becomes very short, the analog-to-digital converter may not be able to register enough points to define the input properly.

The output signal from the comparator, v_B, is often utilized in digital systems because its period is directly proportional to the amplitude of the input signal. If this type of analog-to-pulse-width converter is desired, the AND gate and clock-pulse generator can be eliminated.

4.5 APPLICATION HINTS

Stray coupling between the output and the offset-adjustment terminals in voltage comparators can occasionally cause oscillations. During circuit layout every attempt should be made to keep these leads separated as much as possible. If the offset-adjustment terminals are not utilized, it is advisable to tie these pins together to minimize the effect of the feedback. If the offset adjustment is employed, the same result can be accomplished by connecting a fairly large (typically 0.1 μF capacitor between these pins).

When driving the inputs of a comparator from a low-impedance source, a limiting resistance should be placed in series with the input leads to limit the peak current. This is especially important when the inputs are taken outside a piece of equipment where they could accidentally be connected to high-voltage sources. Low-impedance sources do not cause a problem unless their output voltage is larger than either the negative or positive supply voltage. However, since the supply voltages go to zero when they are turned off, isolation is usually needed. An alternative connection for protecting the inputs from excessive current is achieved by placing two diodes between the positive and negative inputs of the comparator. The diodes are placed in parallel in alternate connection such that if the input voltage becomes excessively large, one diode of the pair conducts, preventing excessive current from entering the comparator. Capacitors larger than 0.1 μF connected to the input terminals should be treated as low-impedance sources and isolated with a resistor. A charged capacitor can maintain an input voltage larger than the supply voltage, should the supply be shut off abruptly.

Precautions should be taken to ensure that the power-supply voltages never become reversed, even under transient conditions. With reverse voltages greater than about a volt, the IC can conduct excessive current, which will open circuit internal aluminum interconnects. This usually takes more than 0.15 A. If there is a possibility of power-supply reversal, clamping diodes with an adequate peak-current capability should be installed across the power-supply bus lines.

4.6 CONCLUSIONS

We have described the operation of an ideal voltage comparator together with several important practical limitations to ideal VC operation. The limitations were traced to selected performance parameters of the individual circuit blocks comprising the comparator. We discussed the definitions of typical parameters for the HA-2111 voltage comparator and interpreted them for purposes of device application. Several useful circuits for accomplishing specific applications were given, and the operation of each was described. In addition, we suggested ways of modifying the circuits to suit other applications as may be required.

BIBLIOGRAPHY

1. Angelo, E. James, Jr., *Electronics: BJTs, FETs, and Microcircuits*, McGraw-Hill, New York, 1969

2. Barna, Arpad, *Operational Amplifiers*, Wiley, New York, 1971.

3. Belove, Charles, Harry Schachter, and Donald L. Schilling, *Digital and Analog Systems, Circuits, and Devices: An Introduction*, McGraw-Hill, New York, 1973.

4. Deboo, Gordon J., and Clifford N. Burros, *Integrated Circuits and Semiconductor Devices: Theory and Application*, McGraw-Hill, New York, 1971.

5. Fitchen, Franklin C., *Electronic Integrated Circuits and Systems*, Van Nostrand-Reinhold, New York, 1970.

6. Graeme, Jerald G., Gene E. Tobey, and Lawrence P. Huelsman, *Operational Amplifiers: Design and Applications*, McGraw-Hill, New York, 1971.

7. Graeme, Jerald G. *Applications of Operational Amplifiers: Third Generation Techniques*, McGraw-Hill, New York, 1973.

8. Grebene, Alan B., *Analog Integrated Circuit Design*, Van Nostrand-Reinhold, New York, 1972.

9. Meyer, Charles S., David K. Lynn, and Douglas J. Hamilton, *Analysis and Design of Integrated Circuits*, McGraw-Hill, New York, 1968.

10. Melen, Roger, and Harry Garland, *Understanding IC Operational Amplifiers*, Sams, Indianapolis, 1971.

11. Millman, Jacob, and Christos C. Halkias, *Electronics Devices and Circuits*, McGraw-Hill, New York, 1967.

12. Millman, J., and C. C. Halkias, *Integrated Electronics: Analog and Digital Circuits and Systems*, McGraw-Hill, New York, 1972.

13. Millman, J., and Herbert Taub, *Pulse, Digital, and Switching Waveforms*, McGraw-Hill, New York, 1965.

14. Morris, Robert L., and John R. Miller, *Designing With TTL Integrated Circuits*, McGraw-Hill, New York, 1971.
15. Motorola Semiconductor Products Inc., "Linear Integrated Circuits Data Book," Motorola, Inc., December 1972.
16. Motorola Semiconductor Products Inc., "TTL Integrated Circuits Data Book," Motorola, Inc., May 1972.
17. Peatman, John B., *The Design of Digital Systems*, McGraw-Hill, New York, 1972.
18. Pierce, J. F., and T. J. Paulus, *Applied Electronics*, Merrill, Columbus, Ohio, 1972.
19. RCA Inc., "Linear Integrated Circuits," RCA Corp., Technical Series IC-42, 1970.
20. RCA Inc., "Linear Integrated Circuits and MOS Devices," RCA Corp. No. SSD-202, 1972.
21. Ryder, John D., *Electronic Fundamentals and Applications*, Prentice-Hall, Englewood Cliffs, N.J., 1970.
22. Schilling, Donald L., and Charles Belove, *Electronic Circuits: Discrete and Integrated.* McGraw-Hill, New York, 1968.
23. Stewart, Harry E., *Engineering Electronics*, Allyn & Bacon, Boston, 1969
24. Strauss, Leonard, *Wave Generation and Shaping*, McGraw-Hill, New York, 1970.

CHAPTER 5

OPERATIONAL AMPLIFIERS

The operational amplifier (op amp) is a high-gain, dc-coupled amplifier used principally as the active device in a feedback circuit. Given the sufficient voltage gain of the operational amplifier itself, the feed-forward characteristic of the device and the feedback-loop combination can be tailored to be a function only of the feedback loop and not of the amplifier. The availability of mass-produced, low-cost, monolithic op amps has generated designs of unprecedented precision, speed, reliability, and reproducibility.

The term "operational amplifier" originally designated a specific type of amplifier used in analog computers to perform purely mathematical operations, such as summation, subtraction, integration, and differentiation. The versality afforded by the op amp has now established it as a basic building block in virtually every phase of electronics. Op amps serve as signal conditioners, power regulators, active filters, function generators, analog-to-digital and digital-to-analog converters, and in many other applications.

Our first intention in this chapter is to identify the operational characteristics of an ideal op amp. Then we discuss the definitions of performance parameters commonly given on data sheets of practical op amps. Many applications of op amps are considered, and criteria are developed for utilizing feedback to accomplish specific system requirements. Specifically, the applications of negative-feedback configurations are discussed for generating voltage, current, and instrumentation amplifiers, and voltage sources, signal summers, integrators, differentiators, logarithmic generators, and active filters. Positive-feedback principles are applied to develop circuits for multivibrators, sine-wave oscillators, and triangular-wave generators.

In the latter part of the chapter we examine the structure of typical op amps assembled from the circuit blocks presented in Chapter 3. We will see how practical limitations arising in and from interconnection of these blocks restrict the ideal operation of the amplifier applications considered previously. Selected measurement schemes for obtaining numerical evaluations of the op amp performance parameters are presented, and finally we set forth some generalized techniques useful for optimizing op amp performance in the applications already considered.

5.1 IDEAL OPERATIONAL AMPLIFIERS

The ideal properties of an op amp are tabulated in Figure 5.1, together with ranges of typical specifications found in various device data sheets.

Stage Characteristic	Performance Parameters	Ideal	Typical
Input characteristics	Offset voltage	0 V	0.5–5 mV
	Offset voltage drift with temperature	0 V/°C	1–50 μV/°C
	Bias current	0 A	1 nA–100 μA
	Offset current	0 A	1 nA–10 μA
	Input resistance[a]	$\infty \Omega$	10 kΩ–1000 MΩ
	Common-mode range[b]	Full power-supply range	1–5 V less than power-supply limits
	Maximum differential-input voltage[b]	Difference between power-supply voltages	one-quarter to full power-supply difference voltage
Transfer characteristics	Large-signal voltage gain	$\rightarrow \infty$ V/V	10^3–10^6 V/V
	Common-mode rejection ratio	∞ dB	60–120 dB
	Unity-gain bandwidth	∞ Hz	1–100 MHz
Output characteristics	Output voltage swing[b]	Full power-supply range	1–5 V less than power-supply limits
	Output current	Capability of power supply	1–30 mA
	Full-power bandwidth	∞ Hz	10 kHz–2 MHz
Transient response	Rise time[c]	0 s	10 ns–10 μs
	Slew rate	∞ V/s	0.1–100 V/μs
Power-supply characteristics	Supply current	0 A	0.05–25 mA
	Power-supply rejection ratio	∞ dB	60–100 dB

[a] Highly dependent on input stage. With FET front-end, $10^{10} \leqslant R_i \leqslant 10^{12}$.
[b] The upper limit is set by the power-supply voltages.
[c] Highly dependent on external loading conditions and frequency compensation employed.

Figure 5.1 Ideal and typical op amp specifications.

The typical performance parameters are given for comparative purposes and are useful as a reference guide for comparing new developments of operational amplifiers. (The selling price of an op amp, often an extremely important parameter to a design engineer, seldom appears on a specification sheet. Of course the ideal op amp would be free. However, practical op amps typically sell for prices between 40¢ and $20.00 each.) The meaning and significance of the various device parameters on circuit operation are treated in the next section. Although the parameters of an ideal op amp can never be realized, its characteristics do serve as a guide for comparing present state-of-the-art devices.

5.2 DEFINITIONS OF OP AMP PARAMETERS

Let us now consider working definitions of the parameters presented in Figure 5.1

Input-Offset Voltage (V_{IO}) The differential input voltage required to produce zero output voltage.

Offset-Voltage Drift with Temperature or *Temperature Coefficient of Offset Voltage* (T.C. V_{IO}) The ratio of the average change in input-offset voltage to a change in ambient temperature.

Input-Bias Current (I_B) The current flowing into (or out of) either input terminal while the output voltage is near zero volts. Usually expressed as the average of the two input currents, that is,

$$I_B = \frac{|I_{in}(+)| + |I_{in}(-)|}{2} \tag{5.1}$$

Input-Offset Current (I_{OS}) The difference between the two input currents, that is,

$$I_{OS} = |I_{in}(+) - I_{in}(-)| \tag{5.2}$$

Input Resistance (R_{in}) The ratio of a small change in the differential input voltage to a resulting change in the input current, with the output voltage remaining in its linear region, that is,

$$R_{in} = \left| \frac{\Delta V_D}{\Delta I_{in}} \right| \tag{5.3}$$

Another parameter, the *common-mode input resistance* is usually much higher and is the ratio of a change in common-mode voltage to a resulting change in input-bias current.

Common-Mode Range (CMR)   The range of common-mode input voltage (a voltage measured with respect to ground and applied simultaneously to both inputs) over which the amplifier will operate satisfactorily (usually allowing a 6 dB degradation of CMRR from the minimum specified).

Maximum Differential Input Voltage   The V_D which, if exceeded, may cause damage to the device.

Large-Signal Voltage Gain (A_{vo}) The ratio of a change in output voltage to the change in differential input voltage causing it:

$$A_{vo} = \frac{\Delta V_{out}}{\Delta V_D}$$

(5.4)

Also known as the *dc open-loop gain.*

Common-Mode Rejection Ratio (CMRR)   The ratio of common-mode input voltage change to a change in input offset voltage created by it, usually expressed in decibels:

$$CMRR = 20 \log \frac{\Delta V_{CM}}{\Delta V_{IO}}$$

(5.5)

Unity-Gain Bandwidth (BW)   The frequency at which the open-loop gain is zero dB. Another term, gain–bandwidth product (GBW) is the product of the available open-loop gain at a certain frequency times that frequency. In an amplifier with a −6 dB per octave open-loop gain slope extending through 0 dB, GBW = BW.

Output-Voltage Swing (V_{out}) The maximum output voltage available under specified loading conditions.

Output Current (I_{out}) The maximum output current available, usually measured at $\pm V_{out}$.

Full-Power Bandwidth (FPBW)   The maximum frequency at which an undistorted sine-wave output with $\pm V_{out}$ peak-to-peak may be amplified. Usually limited by slew rate.

Rise Time (RT)   Measured from the 10 to 90% point of the leading edge of the output wave form when the input is subjected to a small-signal voltage pulse.

Slew Rate (SR)   The ratio of the output-voltage swing, measured from the 10 to 90% point of the leading or trailing edge, to the time required for the output to traverse this level, measured under large-signal conditions.

Supply Current (I_S) The quiescent supply current required by the amplifier, measured when the output is at zero volts so that no current is delivered to the load.

Power-Supply Rejection Ratio (PSRR)   The ratio of a change in power-supply voltage to a resulting change in input-offset voltage, usually expressed in decibels, that is,

$$PSRR = 20 \log \frac{\Delta V_S}{\Delta V_{IO}} \qquad (5.6)$$

Since the PSRR may be different for a change in each supply, it is usually specified for the worst-case condition.

We use these parameter definitions in many of the sections that follow. Now let us consider a number of useful applications of operational amplifiers.

5.3   APPLICATIONS OF OPERATIONAL AMPLIFIERS

Inverting Amplifier

To better understand the operation of negative-feedback networks, let us look for a moment at the op amp by itself. Figure 5.2 illustrates an op amp

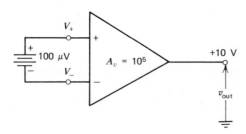

Figure 5.2   Open-loop operational amplifier.

that is "ideal" except that it has a finite open-loop, dc voltage gain, A_v. If we wish to make the output voltage assume a level, v_{out}, we must apply a differential input voltage $(V_+ - V_-)$ equal to v_{out}/A_v. For example, if $A_v = 100,000$ and the desired $v_{out} = +10$ V, we must apply

$$V_+ - V_- = \frac{v_{out}}{A_v} = \frac{+10 \text{ V}}{100,000} = +100 \text{ } \mu\text{V} \approx 0 \qquad (5.7)$$

between the inputs. We can see that if A_v is very large, the differential input voltage is almost negligible compared with v_{out}. This leads to a simple and very useful analysis tool:

[1]   *If a linear relationship is maintained between the input and output voltages in an op amp, we may assume for a first-order approximation that the differential input voltage is virtually zero.*

Now let us examine a simple negative-feedback network, the inverting voltage amplifier illustrated in Figure 5.3. For the moment, we assume that the circuit is stable and that the output voltage v_{out} is some function of the input-signal voltage v_{in}.

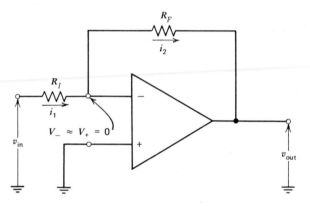

Figure 5.3   Inverting amplifier.

To satisfy the approximation in [1], v_{out} must always assume a level, for any value of v_{in}, to make $V_- = 0$ V. We can see that if v_{in} is positive, v_{out} must be negative to maintain V_- at zero volts. Conversely, if v_{in} is negative, v_{out} must be positive. If no current flows into the negative input of the op amp, then $i_1 = i_2$; and if $V_- = 0$, we can write

$$\frac{v_{in}}{R_I} = -\frac{v_{out}}{R_F} \tag{5.8}$$

By rearranging, we have

$$G_- = \frac{v_{out}}{v_{in}} = -\frac{R_F}{R_I} \tag{5.9}$$

where G_- is the closed-loop voltage gain of the amplifier.

To demonstrate that the circuit is stable for dc conditions, we show that any arbitrary change in v_{out} will generate an error signal to counteract that change. Suppose for some reason v_{out} tries to change in the positive direction while v_{in} remains constant. This will produce a positive change in V_-, which would try to drive v_{out} in the negative direction until V_- again approaches zero volts. Since a tendency for a negative change in v_{out} will be corrected in a corresponding manner, the connection is at least dc stable.

Equation 5.9 shows that the circuit gain is determined by the ratio of two external resistors and illustrates a very important characteristic of most op amp applications:

[2]   *Closed-loop circuit performance is determined mainly by the component values in the feedback network if op amp gain A_v is very high and other op amp error sources are very low.*

This is a powerful tool to the systems designer because the design can tolerate large variations in op amp parameters, such as open-loop gain, with negligible effect on closed-loop performance.

Noninverting Amplifier

A circuit connection in which the output signal has the same polarity as the input signal appears in Figure 5.4a. Figure 5.4b shows the similarity between the inverting and noninverting amplifier connections. Note that the input signal and ground connections are simply reversed. In Figure 5.4c we have redrawn the noninverting configuration to show the voltage-divider function of the feedback network. Using voltage division, the voltage fed back, V_-, is

$$V_- = v_{out} \frac{R_I}{R_F + R_I} \tag{5.10}$$

From Equation 5.7, we know that

$$V_- \approx V_+ = v_{in} \tag{5.11}$$

Therefore, we can write

$$G_+ = \frac{v_{out}}{v_{in}} \approx \frac{R_I + R_F}{R_I} = 1 + \frac{R_F}{R_I} \tag{5.12}$$

where G_+ is the closed-loop voltage gain of the noninverting configuration.

An important consideration in this type of feedback connection is that the voltage fed back, V_-, tends to be very nearly equal to the input voltage v_{in}. Also; v_{out} will assume whatever level is necessary to force $V_- = v_{in}$.

Note the subtle difference in the closed-loop gain expression for the

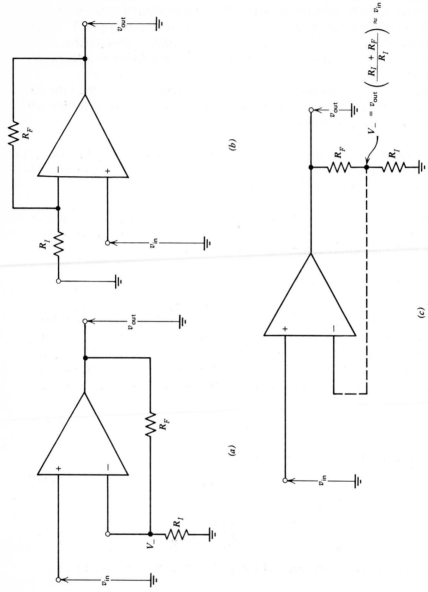

Figure 5.4 The noninverting amplifier. (*a*) Circuit configuration. (*b*) Rearranged circuit configuration to show similarity to inverting amplifier. (*c*) Generation of feedback signal by voltage-divider network at output.

inverting amplifier (G_- of Equation 5.9) compared with that for the noninverting amplifier, (G_+ of Equation 5.12). This is often surprising, since the two connections are identical except for the point of input-signal introduction. The difference arises because in the inverting connection, R_I and R_F form a voltage divider both for the input signal and the feedback signal, whereas in the noninverting connection the voltage divider acts only on the feedback signal. Therefore, for circuits having identical resistor ratios, the closed-loop gain of the noninverting circuit will be greater in magnitude by one than that of the inverting circuit. However, in computing closed-loop frequency response, and in computing effects at the output of noise, offset voltage, and current, and so on, the gain factor, $1 + R_F/R_I$, must be used for both the inverting and noninverting connections.

There are some related differences in the performance of the two feedback connections. In the inverting connection, R_I feeds a virtual ground at the negative input; thus the circuit input impedance is equal to R_I. In the noninverting connection the circuit input impedance is equal to the op amp common-mode input impedance, which is typically many megohms. In the noninverting connection, since v_{in} appears almost equally on both op amp inputs, v_{in} is a common-mode input signal that will produce an additional error:

$$v_{out(error)} = \frac{v_{in}}{CMRR}\left(1 + \frac{R_F}{R_I}\right) \tag{5.13}$$

where CMRR is expressed as a numerical ratio (e.g., 80 dB = 10,000). Thus the op amp chosen for noninverting applications should have a common-mode range equal to or greater than the input signal range, as well as a high CMRR.

Figure 5.5 shows a special case of the noninverting amplifier connection where $R_F = 0$ and $R_I = \infty$, reducing the closed-loop gain G_+ to 1. This is a useful circuit for impedance conversion in connecting a high-impedance

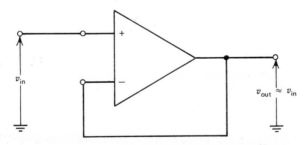

Figure 5.5 Voltage-follower op amp configuration.

source signal to a low-impedance load. Usually called a "voltage follower," it performs the same function as a single-stage cathode or emitter
follower with much lower dc errors.

Differential Amplifier

A differential input amplifier combining the inverting and the noninverting
connections (Figure 5.6) is easily analyzed by considering the effect of
each input separately. If the v_2 input is grounded, the circuit is an
inverting amplifier for signals at the v_1 input with a G_- gain of $-R_F/R_I$. If

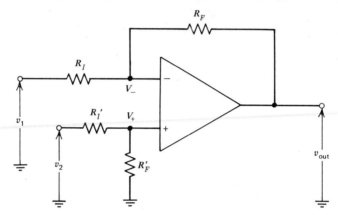

Figure 5.6 Op amp used as a differential amplifier.

the v_1 input is grounded, the circuit becomes a noninverting amplifier with
R_F' and R_I' forming a voltage divider at the input; thus the voltage at the
(+) terminal is

$$V_+ = v_2 \frac{R_F'}{R_I' + R_F'} \tag{5.14}$$

and

$$v_{\text{out}} = V_+ \frac{R_I + R_F}{R_I} \tag{5.15}$$

If $R_I = R_I'$ and $R_F = R_F'$, the gain at the v_2 input is $+R_F/R_I$. The complete
expression for the amplifier is then

$$v_{\text{out}} = (v_2 - v_1) \frac{R_F}{R_I} \tag{5.16}$$

This means that ideally the output responds only to the difference in
voltage between the two inputs and does not respond to a voltage

common to the two inputs (common-mode voltage). In other words, the same output should occur when the inputs are +2.0 and +2.1 V as when the inputs are 0 and +0.1 V, respectively. This circuit is particularly useful for amplifying signals that have been transmitted in differential form on balanced transmission lines, since most of the noise pickup will be common-mode voltage. For best CMRR, the ratio $R_F/R_I = R_F'/R_I'$ must be held as close as possible, since any mismatch will cause a portion of the common-mode signal to appear differentially across V_+ and V_-. The error due to op amp CMRR is computed from

$$v_{\text{out(error)}} = \frac{V_{CM}}{\text{CMRR}}\left(\frac{R_F}{R_I}\right) \tag{5.17}$$

Errors due to noise, offset voltage, and other causes must be multiplied by $(1 + R_F/R_I)$ to compute the output error.

Summing Amplifier

Figure 5.7 illustrates a further modification of the differential amplifier in which a number of input signals may be algebraically added and/or

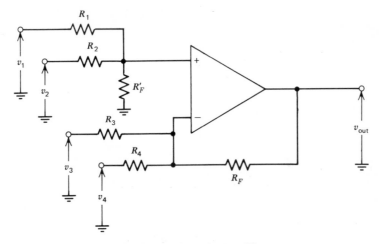

Figure 5.7 Summing amplifier.

subtracted from one another. The equation for this circuit function is

$$v_{\text{out}} = \frac{R_F' \| R_2}{R_1 + (R_F' \| R_2)}\left(1 + \frac{R_F}{R_3 \| R_4}\right)v_1 + \frac{R_F' \| R_1}{R_2 + (R_F' \| R_1)}$$

$$\times \left(1 + \frac{R_F}{R_3 \| R_4}\right)v_2 - \frac{R_F}{R_3}v_3 - \frac{R_F}{R_4}v_4 \tag{5.18}$$

Any number of inputs may be added to either side of the amplifier with appropriate gain-setting resistors, or all inputs may go to one side of the amplifier to form a noninverting or inverting summation circuit. In any case, with $R_F = R_F'$, it is essential that the parallel combination of resistors at the negative input equal the parallel combination of resistors at the positive input. This can be accomplished by adding a resistor to ground from the appropriate input to achieve identical resistance levels to ground at both inputs.

This circuit is useful in analog computation and in any system in which several analog signals must be added and/or subtracted.

Current-To-Voltage Converter

Sometimes it is desirable to produce a circuit with a very low input impedance and an output voltage proportional to the input current. The inverting amplifier circuit accomplishes this function by making $R_I = 0$.

As Figure 5.8 indicates, the input current from the source must pass through the feedback resistor R_F. To maintain the differential input

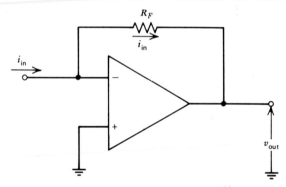

Figure 5.8 Current-to-voltage converter.

voltage close to zero, the output voltage must assume a level

$$v_{out} = -i_{in}R_F \qquad (5.19)$$

Therefore, the output voltage is directly proportional to the input current.

The current-to-voltage converter is especially useful as an ammeter circuit with 0 Ω shunt resistance, and as an amplifier for signals produced by high-impedance current sources. Figure 5.9 reveals one of the advantages of the current-to-voltage configuration as a high-speed photodiode amplifier. The noninverting voltage amplifier circuit in Figure 5.9a is

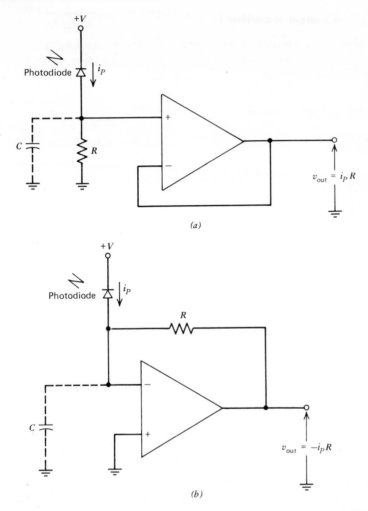

Figure 5.9 Photodiode amplifier using (a) the voltage-follower connection, (b) the inverting amplifier configuration.

suitable only for low-frequency operation. Since only a few micro-amperes photocurrent is generated by the diode, R must be very large, and the stray capacitance at the amplifier's positive input will limit the frequency response. The current-to-voltage amplifier circuit in Figure 5.9b has a much better frequency response, since the voltage at the negative input is always at virtual ground, and the stray capacitance does not need to be charged or discharged with changes in photocurrent.

Instrumentation Amplifier

In analog instrumentation, the signal transducers are frequently located some distance from the measurement system, and their signal levels are often low and their source impedances high. A general-purpose amplifier for conditioning these signals should have differential inputs, high input impedance, high common-mode rejection, and simple gain adjustment. A circuit containing three op amps (Figure 5.10) meets all these requirements. The circuit has two gain stages. The first stage consists of A_1, A_2,

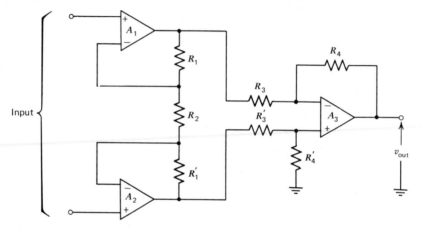

Figure 5.10 Instrumentation amplifier with three op amps.

R_1, R_2, and R_1'. The second stage, composed of A_3, R_3, R_4, R_3', and R_4' represents a standard differential amplifier configuration, as discussed previously.

Looking at the first stage alone we can analyze its performance with a differential input, as in Figure 5.11a. Since the differential input voltage of each op amp is nearly zero, we know that the voltage appearing on each side of R_2 determines the current through R_2. Since this current also flows through R_1 and R_1', we can calculate the voltage appearing at each output. The differential gain of this stage is given by

$$\frac{\Delta V_{\text{out}}}{\Delta V_{\text{in}}} = \left(1 + \frac{R_1}{\frac{1}{2}R_2}\right) \tag{5.20}$$

Since for differential input signals there is a virtual ground at the midpoint of R_2, this is analogous to the noninverting amplifier circuit gain.

Figure 5.11b shows the same circuit with a common-mode voltage input. Since there is no voltage drop across R_2, no current flows through

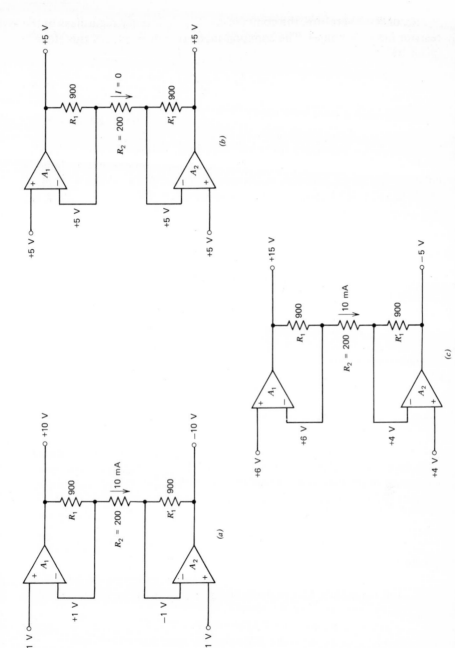

Figure 5.11 Instrumentation amplifiers for (*a*) differential-mode input, (*b*) common-mode input, and (*c*) differential- and common-mode inputs.

155

R_1, R_2, or R'_1. Therefore, the common-mode gain is unity regardless of the resistor network values. The common-mode rejection ratio of this stage is given by

$$CMRR = \left| \frac{A_d}{A_c} \right| \tag{5.21}$$

where A_d and A_c are the difference and common-mode gains, respectively.

Figure 5.11c illustrates the stage with both differential and common-mode signals applied and shows that the output common-mode signal is equal to the input common-mode signal (+5 V). Most of this common-mode signal will be rejected in the second stage. The overall gain of the two-stage cascade is

$$\frac{\Delta V_{out}}{\Delta V_{in}} = \left(1 + \frac{2R_1}{R_2}\right)\frac{R_4}{R_3} \tag{5.22}$$

The gain may be easily adjusted without disturbing circuit symmetry by varying the resistance of R_2.

The first stage may also be used by itself when a circuit with both differential inputs and differential outputs is required.

Analog Integrator

An integrator circuit like the one in Figure 5.12 is quite similar to the inverting amplifier except that a capacitor has been substituted for the feedback resistor. Like the inverting amplifier, the negative input of the op amp will tend to remain at virtual ground, which means that the condition i_2 must equal i_1. However, current can flow through a capacitor only when the voltage across it changes. In the integrator, the *rate of change* of output voltage is proportional to the input voltage. In calculus notation this is expressed as

$$v_{out} = -\frac{1}{RC} \int_0^T v_{in} \, dt \tag{5.23}$$

Figure 5.12b illustrates the integrator's output as a function of time for a step change in the input voltage with an initial condition of zero volts across the capacitor. The rate of change of v_{out} may be expressed as

$$\frac{\Delta V_{out}}{\Delta T} = -\frac{v_{in}}{RC} \tag{5.24}$$

Note that the output voltage does not return to zero if the input goes to zero; the output simply stops moving. Therefore, the output voltage at

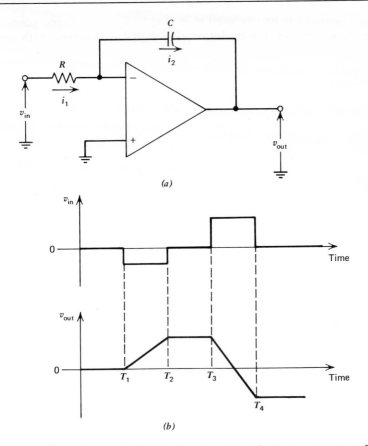

Figure 5.12 Op amp integrator. (*a*) Circuit configuration. (*b*) Input–output wave forms.

any instant is determined by the past history of the input voltage. The output voltage is actually proportional to the area enclosed by the input wave form from $t = 0$ to some time instant T.

When a sinusoidal input is applied to the integrator, a low-pass filter is formed whose gain is inversely proportional to frequency.

In practical circuits, the output voltage will tend to continually drift with zero volts input, eventually reaching one of the amplifier output-voltage limits. This effect is due to the offset voltage and bias current of the amplifier, which act in the same manner as a small input signal. It is usually desirable to periodically discharge the capacitor, either by reversing the input-signal polarity or by shorting the capacitor momentarily. Shunting the capacitor with a high-value resistor will aid the dc stability with some loss in functional accuracy.

The integrator is widely used in analog computers, sweep generators, timers, function generators, integrating digital voltmeters, A/D converters, and active filters.

Differentiator

Figure 5.13 shows a somewhat less widely used circuit, the differentiator. The resistor and capacitor positions are simply reversed from the integrator connection.

As illustrated in Figure 5.13b, the output voltage is now proportional to

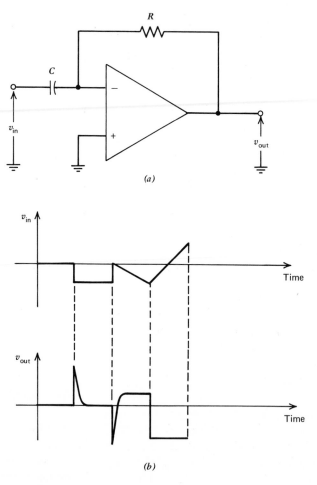

Figure 5.13   Op amp differentiator. (a) Circuit configuration. (b) Input–output wave forms.

the rate of change of the input voltage:

$$v_{out} = -RC \frac{dv_{in}}{dt} \qquad (5.25)$$

With sinusoidal inputs, the differentiator acts like a high-pass filter whose gain is directly proportional to the input frequency. One disadvantage of the differentiator is that the high-pass function accentuates noise at the output.

Logarithmic Amplifiers

Nonlinear functions can be generated by the use of nonlinear resistors, such as diodes and varistors, in the network connected to an op amp.

The diode $V-I$ characteristic (Figure 5.14) indicates a very high

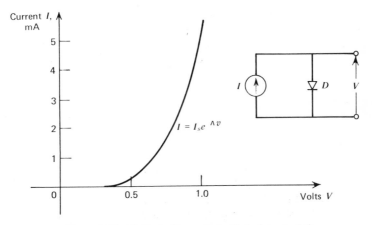

Figure 5.14 Typical silicon diode $V-I$ characteristic.

dynamic resistance at low-currents, and the resistance decreases as current is increased. With selected diodes, this characteristic closely approximates an exponential function over a five-decade range of currents.

If the diode is inserted in an inverting amplifier in place of the feedback resistor (Figure 5.15), a circuit is formed that has high gain at low output levels and low gain at high output levels. This amplifier connection approximates the logarithmic function:

$$v_{out} = -\frac{1}{\Lambda} \ln \frac{v_{in}}{I_s R} \qquad (5.26)$$

where I_s and Λ are the diode parameters discussed in Chapter 3.

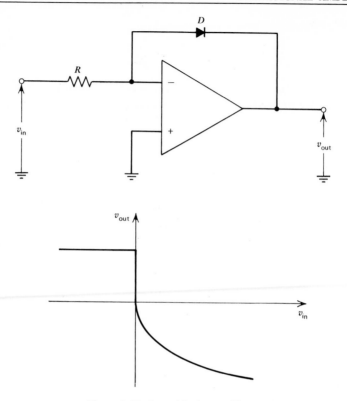

Figure 5.15   Logarithmic amplifier.

Eliminating the input resistor R as in Figure 5.16 creates a current-to-voltage amplifier with the log characteristic

$$v_{\text{out}} = -\frac{1}{\Lambda} \ln \frac{i_{\text{in}}}{I_s} \tag{5.27}$$

Interchanging the diode and resistor as in Figure 5.17 produces an antilog circuit with an input–output characteristic given by

$$v_{\text{out}} = -RI_s e^{\Lambda v_{\text{in}}} \tag{5.28}$$

Selected transistors are often substituted for the diode as in Figure 5.18 to extend the range of logarithmic approximation.

The logarithmic amplifier is a unipolar circuit requiring the input voltage to be positive at all times. Otherwise the diode becomes reverse biased, and logarithmic operation ceases. Since the gain is very high for low-level input signals, offset voltage and bias current of the op amp must be very low.

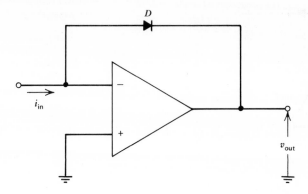

Figure 5.16 Logarithmic current-to-voltage converter.

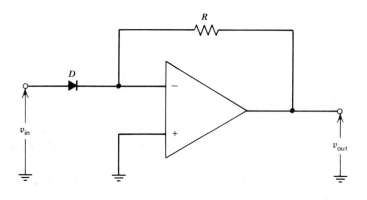

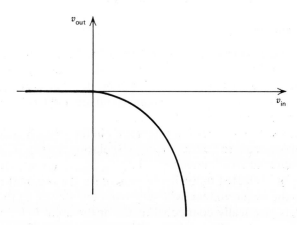

Figure 5.17 Antilog amplifier.

161

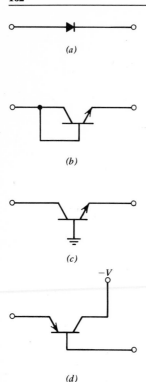

(a)

(b)

(c)

(d)

Figure 5.18 Transistors commonly substituted for diode in the log amplifiers to extend the useful operating range.

The log amp may be used for scale-factor conversion (changing volts to decibels, etc.); in combinations for analog multiplication or division; and as audio signal compandors for improvement of signal-to-noise ratios.

Voltage Regulators

A precise voltage source is required in A/D, D/A, and many other systems. Temperature-compensated zener diodes now have temperature stabilities better than 0.0005%/°C. An op amp is often required for use with the zener diode to translate the zener voltage and to allow appreciable current to be drawn from the reference.

Figure 5.19a shows a precision reference circuit—simply a noninverting amplifier connected to a zener diode. The biasing resistor R_3 should be chosen to produce the recommended operating zener current. The op amp gain resistors are selected individually to achieve the required reference output, since the zener voltage will vary from one device to the next. If the top of R_1 is physically connected to V_{ref} at the point to be regulated, any voltage drops along the V_{ref} line will be compensated. The circuit can

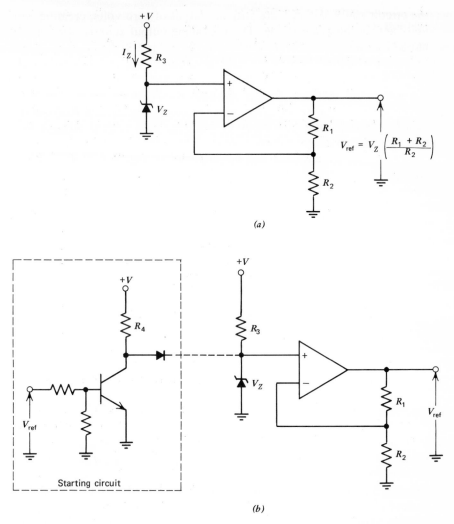

Figure 5.19 Precision voltage regulators. (a) Basic circuit. (b) A self-biased regulator.

easily be changed to a negative reference by reversing the zener diode and connecting R_3 to a negative supply.

For best precision, the circuit in Figure 5.19a should have a well-regulated supply connected to the top of R_3. If that is not available, a self-biased reference, such as in Figure 5.19b may be constructed.

The output may have more than one stable condition, and this problem is frequently overlooked in regulator design. For example, in Figure 5.19b the circuit could also operate with an output of zero volts, creating zero volts across the zener diode. The prevailing output state is often determined by transients generated when the power is initially applied—a less than sure situation. Adding the starting circuit in Figure 5.19b eliminates the undesirable second state. If V_{ref} is zero, the transistor is turned off, and the zener receives current through R_4. As soon as V_{ref} becomes positive, the transistor turns on, and the diode disconnects R_4 from the zener; thus the zener now receives well-regulated current through R_3.

When more output current is required than the op amp can furnish, a transistor may be added to the op amp output. Figure 5.20 illustrates a

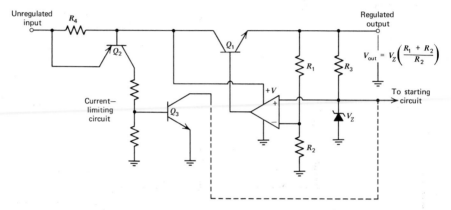

Figure 5.20   Precision voltage regulator with output short-circuit protection.

voltage regulator with up to several amperes output capability, depending on the choice of Q_1. An additional circuit consisting of R_4, Q_2, and Q_3 has been added to protect the power transistor, Q_1, from accidental burnout in the event of an overloaded or shorted output. The resistor R_4 is selected so that its voltage drop equals V_{BE} (about 0.7 V at +25°C) when the maximum allowable output current flows. If the output current tends to exceed this value, Q_2 and Q_3 will turn on, reducing the output voltage toward zero; thus a constant output current is maintained until the overload is removed.

5.4   POSITIVE–FEEDBACK APPLICATIONS OF OPERATIONAL AMPLIFIERS

The op amp applications considered in the previous section used feedback techniques either to improve the open-loop performance of the amplifier or to produce some specialized relationship between the input and output amplifier signals. In each application we examined, the output signal was fed back to the inverting terminal of the op amp in a way that served to reduce the input signal level. Such feedback is called inverse, degenerative, or negative feedback.

If we change our feedback connection so that the output is coupled to the noninverting input of the op amp, the feedback will tend to increase the input signal. This phenomenon is called regenerative or positive feedback. Among the many useful applications possible with positive-feedback circuits are sinusoidal oscillators, switching circuits, and function generators. Let us now examine the operational properties and application of several of these circuits.

Square-Wave Generator

The square-wave generator* or clock is a computer circuit used to produce a timing signal for operating the various parts of an entire computer system. A square-wave generator acts as a self-triggered switch to continuously alternate back and forth between two dc voltage levels without utilizing any outside triggering signal. A simplified arrangement for this function appears in Figure 5.21. The frequency of oscillation is $1/T$.

A square-wave generator that uses an op amp as the switching device is presented in Figure 5.22. The amplitudes of the output voltage are

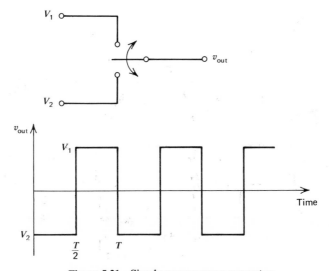

Figure 5.21 Simple square-wave generator.

* Also called an astable multivibrator.

controlled by the $+V_{CC}$ and $-V_{CC}$ power supplies. The output-voltage wave form has sharp leading and trailing edges, typically making the transition from one saturation level to the other in a few microseconds. This transition time is determined mainly by the slew rate of the op amp according to

$$t_{slew} \approx \frac{2V_{CC}}{SR} \tag{5.29}$$

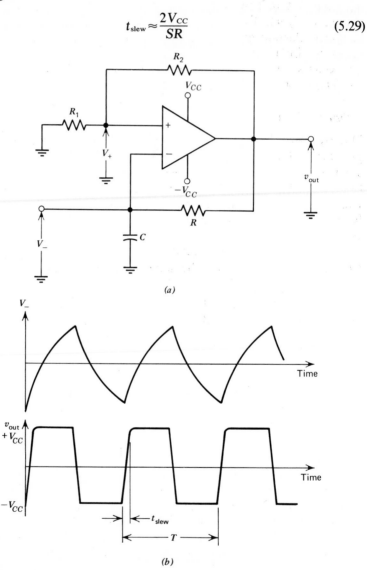

(a)

(b)

Figure 5.22   Square-wave generator. (a) Circuit using an op amp. (b) Circuit wave shapes.

where SR is the slew rate of the op amp. The period of oscillation is T and is given by

$$T = 2RC \ln \left(1 + \frac{2R_1}{R_2}\right) \tag{5.30}$$

where RC is the circuit time constant.

To understand the operation of the square-wave generator, assume that the positive feedback has forced the op amp to saturate in its positive-output state so that $v_{out} \approx +V_{CC}$. Under this condition, the noninverting voltage V_+ will be

$$V_+ = \frac{R_1 V_{CC}}{R_1 + R_2} \tag{5.31}$$

The inverting input voltage V_- will be rising exponentially toward $+V_{CC}$ with a time constant of RC. When $V_+ = V_-$, the circuit will switch from its positive-output state $(+V_{CC})$ to the negative-output state $(-V_{CC})$. In this new state, we have

$$V_+ = -\frac{R_1 V_{CC}}{R_1 + R_2} \tag{5.32}$$

However, the capacitor C keeps V_- from changing its level instantaneously. The circuit remains locked in the negative-output state until V_- can become equal to V_+, as given by Equation 5.32. When $V_+ = V_-$, the output voltage again switches to $+V_{CC}$, and the cycle is repeated. Note the positive feedback utilized around the op amp through R_1 and R_2 to the noninverting input. The combined voltage gain of this network and the amplifier must be greater than unity to ensure proper switching action.

Triangular-Wave Generator

A useful triangular-wave generator can be produced by applying the square-wave output as the input to the integrator circuit shown in Figure 5.12. The integrator's output will be a train of alternating positive and negative ramps of the same period T as the square-wave input. A very versatile square- and triangular-wave-form generator of this type (Figure 5.23) uses three op amps: A_1 produces the square wave, which is then integrated by A_2 to generate the triangular wave shape. The third op amp A_3 functions as a variable-gain output amplifier for buffering A_1 and A_2. The amplitude of the square and triangular wave forms is variable from 0.2 to 20 V peak-to-peak by R_9 and S_3. The rise time of the square wave is less than 450 ns, with rise times less than 200 ns possible with reduced amplitudes.

The square-wave generator consists of a simple hysteresis circuit,

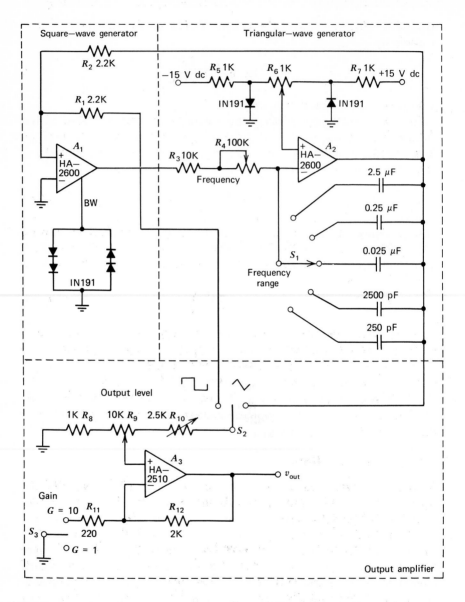

Figure 5.23 Versatile function generator producing square and triangular wave forms.

triggered by the triangular-wave generator. The output voltage of the square-wave generator is clamped to the desired level by diodes connected to the bandwidth control point. The four diodes shown give an output of 2 V peak-to-peak. Two diodes will give 1 V peak-to-peak output and a faster rise time. The ratio of the amplitude of the square wave to the triangular wave is equal to the ratio of R_1 to R_2. The rise time is generally limited by the slew rate of the operational amplifier. The best wave form is obtained by using an HA-2600, but much better rise times can be achieved if amplifiers having very large slew rates are employed. However, these amplifiers may produce undesirable overshoots. The triangular-wave generator consists of an integrator that integrates the output of the square-wave generator. The frequency of the function generator is controlled by the ramp rate of the triangular wave. Switch S_1 selects the integrating capacitor, which changes the frequency range in decade steps. The variable frequency control is R_4. The frequency of the function generator f is given by

$$f = \frac{1}{4(R_3 + R_4)C} \frac{R_1}{R_2} \qquad (5.33)$$

The symmetry of the wave form is adjusted by R_6, which controls the reference voltage of the integrator. The frequency will change if the symmetry is changed by more than approximately 10%. The noninverting input of the op amp can be grounded if it is not necessary to change the symmetry of the wave form. The resulting symmetry is very good. The HA-2600 is chosen because it produces accurate integration, since the typical input-bias current is low: 3 nA. The HA-2510 is preferable if it is necessary to operate the function generator at frequencies as high as 1 MHz.

The output amplifier consists of the simple noninverting connection. The HA-2510 operates well in this stage because of its high slew rate of 50 V/μs. Switch S_3 selects a gain of 1 or 10. The variable output attenuator R_9 sets the input level to the amplifier. Potentiometer R_{10} serves as an output-level calibration control. The maximum output current should be limited to 20 mA. The load impedance should be greater than 600 Ω for a gain of 10 and 50 Ω for unity gain.

Monostable Multivibrator

The monostable multivibrator, or one-shot, is a circuit configuration employing positive feedback and having only one stable state. However, when the circuit is properly triggered, it will leave the stable state and stay in the new "quasi-stable" state for some predetermined time interval; then it will return to the stable state.

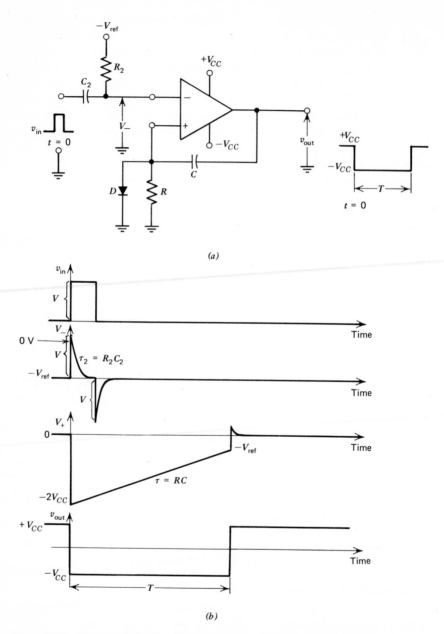

Figure 5.24 "One-shot" circuit using an op amp. (a) Circuit diagram. (b) Wave shapes.

Figure 5.24 shows a basic one-shot circuit using an op amp. Circuit operation can best be understood by first considering the state of the circuit before $t = 0$ when the input pulse is applied. Since the inverting input is connected to a negative reference potential $(-V_{ref})$, v_{out} will be clamped to the positive saturation voltage of the op amp (approximately $+V_{CC}$ volts). When the input pulse v_{in} is applied through the high-pass network of R_2–C_2, the op amp is triggered to its other saturation state, causing v_{out} to immediately go to $-V_{CC}$. The negative-going output pulse is coupled through C_1 to the noninverting input, which keeps V_{out} clamped to $-V_{CC}$. The output remains at $-V_{CC}$ until the potential at the noninverting terminal recovers to $-V_{ref}$. This resetting action occurs after a period of T s as given by

$$T = RC \ln \frac{2V_{CC}}{V_{ref}} \qquad (5.34)$$

Thus the period can easily be controlled by either the RC time constant or the reference voltage. Diode D improves the circuit recovery time by clamping the noninverting voltage to a few hundred millivolts at the end of T when the circuit switches back to its stable state.

Sine-Wave Oscillator

Sine-wave oscillators have long been the "heartbeat" of electronics. They act as signal sources for audio and RF generators. Sinusoidal oscillations also frequently appear as undesirable features in high-gain amplifiers, and to eliminate them we must understand their origin.

Two requirements are essential for producing oscillations in a feedback circuit such as that shown in Figure 5.25a;

1. The total phase shift around the loop must be zero at the single desired frequency of f_o.
2. At f_o the magnitude of the loop gain $|A\beta|$ should be greater than unity for small signal levels and should become exactly unity for the desired output-voltage amplitude.

Because of the phase-shift requirement, all oscillators are called phase-shift oscillators. In practice this term covers oscillators of all types except those which employ negative-resistance devices such as tunnel diodes and unijunction transistors.

When an inverting amplifier is used as the active device in an oscillator circuit, the β feedback network must provide a further phase shift of 180° to meet the first condition of oscillation. The first condition for oscillation is satisfied if the β network has a phase-frequency characteristic as

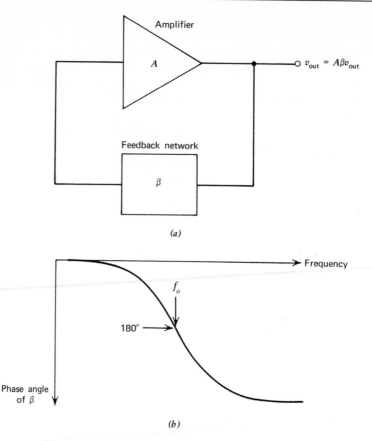

Figure 5.25 Sinusoidal oscillator. (a) Block-diagram organization. (b) A possible phase characteristic for the β feedback network.

shown in Figure 5.25b at a frequency of f_o. Thus it is the β network and not the amplifier A which sets the frequency of oscillation to be very close to f_o. Now let us examine a practical op amp circuit that functions as a sinusoidal oscillator.

The circuit in Figure 5.26a is a Wien bridge oscillator. Positive feedback is provided by the β_1 network, which sets the frequency of oscillation according to

$$f_o = \frac{1}{2\pi\sqrt{R_1 R_2 C_1 C_2}}$$ (5.35)

At f_o, the phase shift of β_1 network is zero degrees. Now note that (connected to) the op amp A is a second feedback network β_2, which we

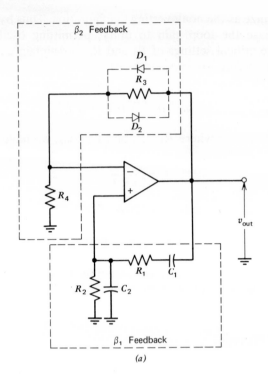

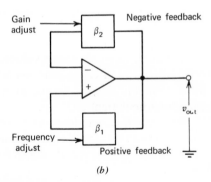

Figure 5.26 The Wien bridge oscillator. (*a*) Complete circuit. (*b*) Simplified equivalent block diagram.

quickly recognize as the noninverting configuration. Thus by adjusting β_2 we can increase the loop gain to unity, permitting oscillation to be sustained. The critical settings of R_3 and R_4 to maintain oscillations are determined according to

$$1 + \frac{R_3}{R_4} \geqslant \frac{R_2 C_1}{R_1 C_1 + R_2 C_2 + R_2 C_1} \tag{5.36}$$

For the simplest case where $R_1 = R_2$ and $C_1 = C_2$, we have

$$f_o = \frac{1}{2\pi R_1 C_1} \tag{5.37}$$

and

$$R_3 \geqslant 2R_4 \tag{5.38}$$

To maintain oscillations with a stable amplitude level, some means of decreasing the amplifier gain with an increase of output signal level must be provided. The two diodes D_1 and D_2 act as bilateral limiters, reducing the effective resistance of R_3 with large output levels.

5.5 ACTIVE FILTERS

Active filters have the greatest advantage over their passive counterparts at frequencies below 10 kHz, where the high-value inductors required in passive filters become bulky, expensive, and less than ideal in performance.

The integrator and differentiator circuits illustrated in Figures 5.12 and 5.13 are simple active filters. Several other examples of the simpler and most useful filter types are treated in this section. Numerous other possible active filter connections can be found in other publications, together with detailed mathematical analyses.

In Figure 5.27, the inverting amplifier and integrator circuits are combined to produce an inverting amplifier with constant gain from dc to a frequency f_2. This break frequency is controlled by the feedback network according to

$$f_2 = \frac{1}{2\pi R_2 C_2} \tag{5.39}$$

Above f_2, the gain decreases by 20 dB/decade (or 6 dB/octave, meaning that the gain decreases by one-half for each doubling of frequency).

In a comparable high-pass active filter connection (Figure 5.28), the

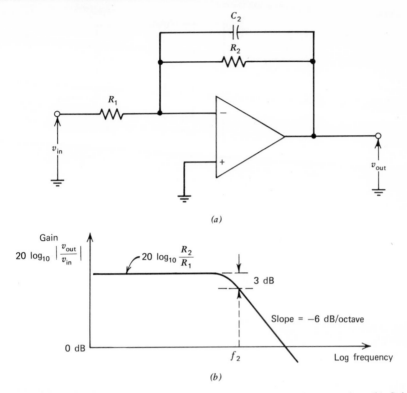

(a)

(b)

Figure 5.27 Single-pole, low-pass active filter. (*a*) Op amp circuit connection. (*b*) Gain versus frequency characteristic.

-3 dB break frequency f_1 is now set by the input network according to

$$f_1 = \frac{1}{2\pi R_1 C_1} \qquad (5.40)$$

If the low-pass and high-pass active filters are combined, an active band-pass filter results (see Figure 5.29).

Active filters may be connected in series to produce steeper "skirt" slopes. In general, cascaded simple filter sections have a lower *sensitivity*; this means that a small change in value of any one component will result in less change to the overall filter response than would have occurred with a comparable single-amplifier complex filter.

A second-order, low-pass active filter connection is diagrammed in Figure 5.30. The transfer characteristic for this configuration is

$$\frac{v_{\text{out}}}{v_{\text{in}}} = -\frac{R_2/R_1}{(1 + sR_2C_2)(1 + sC_1R_1/4)} \qquad (5.41)$$

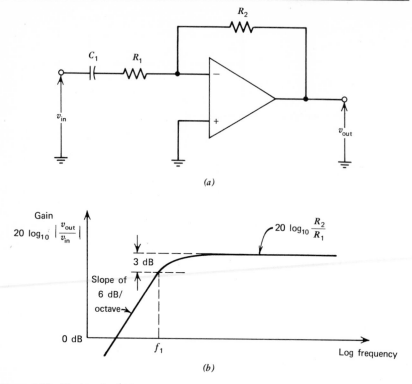

Figure 5.28 Single-pole, high-pass active filter. (*a*) Op amp circuit connection. (*b*) Gain versus frequency characteristic.

Choosing component values according to

$$R_1C_1 = 4R_2C_2 \qquad (5.42)$$

will produce a double pole at

$$f_2 = \frac{1}{2\pi R_2 C_2} \qquad (5.43)$$

Figure 5.31 illustrates a state-variable filter that has gained increasing popularity as integrated multiple op amps have become available. The advantages of this filter are low sensitivity to component variations and the availability of high-pass, low-pass, and band-pass outputs. The filter is made up of a summing amplifier and two integrators wired in a loop. In practice, components are chosen such that

$$R_5C_1 = R_6C_2 \qquad (5.44)$$

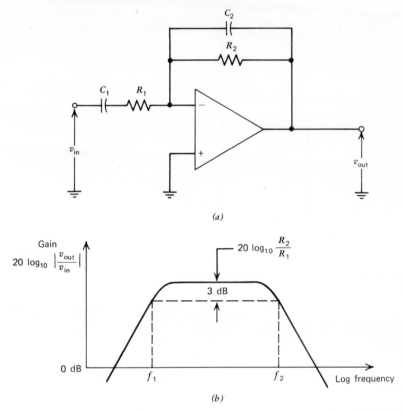

Figure 5.29 Active band-pass filter. (*a*) Circuit connection. (*b*) Frequency characteristic

and the break frequency is

$$f_c = \frac{1}{2\pi R_s C_1} \tag{5.45}$$

for all three outputs. The high-pass and low-pass outputs have 12 dB/octave skirts; and the band-pass output has a sharp response at f_c, with Q determined by the gain-setting resistors of A_1.

5.6 OPERATIONAL AMPLIFIER ORGANIZATION

The block-diagram organization of a typical op amp (Figure 5.32) reveals a close similarity between the voltage comparator and the op amp in that each device typically has a differential amplifier input stage whose characteristics are stabilized against both internal and external influences

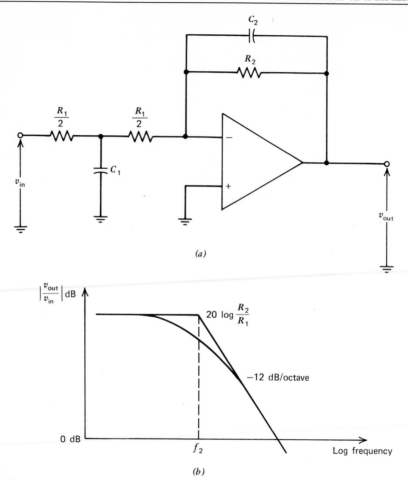

Figure 5.30 Two-pole, low-pass active filter. (a) Circuit configuration. (b) Frequency response.

through the bias network. The input-offset voltage of both the op amp and VC can be adjusted to a minimum level through external networks connected to the offset-adjust pins. The major distinction between the two devices is that op amps usually have highly linear gain stages and output stages, whereas VCs have nonlinear output stages suitable for interfacing with logic circuits.

The aim of most op amp applications is to keep the device in its linear region of operation. For this reason, linear stability is a major consideration when using op amps, although it is not generally encountered with

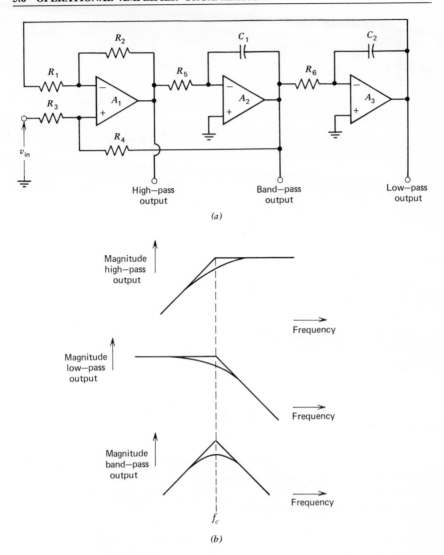

Figure 5.31 State-variable filter. (*a*) Circuit configuration. (*b*) Frequency characteristics of outputs.

VCs. Some form of frequency compensation is usually provided in the gain stage of an op amp. The purpose of a compensation network is either to reduce this signal amplification of high frequencies by the gain stage or to reduce the phase shift existing between the input and output signals. As we discussed in Section 3.7, the gain stage is a logical location for frequency compensation because of its isolation from external loading

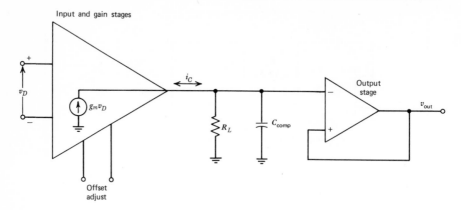

Figure 5.32 Op amp model simplified by combining input and gain stages.

influences. This compensation can be internal, external, or a combination of both. External compensation allows the designer to optimize a particular device's GBW product. Internal compensation has the advantage of simplicity in designs where extended high-frequency performance of the op amp is not a critical factor.

Since linearity is a major concern in op amp operation, the output stage is usually operated class AB. Furthermore to achieve a relatively high output-power level in this stage with low harmonic distortion, a complementary push-pull configuration is used (cf. types discussed in Section 3.8). Some short-circuit protection is usually provided in the output stage to prevent accidental device destruction due to grounding the output or inadvertent connecting of the output to either power supply. In the model shown in Figure 5.32, i_C and v_D are related through an equivalent transconductance g_m, which is the slope of the i_C–v_D curve evaluated at the origin. (Note the similarity between the simplified model and the transconductance amplifier model of Figure 3.15.) Many op amp parameters, such as gain, bandwidth, and slew rate, can be correlated using the simplified op amp model.

5.7 PRACTICAL OPERATIONAL AMPLIFIER LIMITATIONS

Static Response

We will now illustrate the effect of practical device and circuit limitations on the ideal operation of an op amp. Figure 5.33a shows the transfer

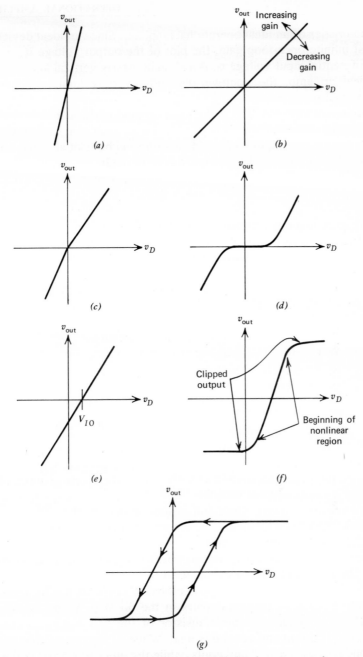

Figure 5.33 Changes in the ideal transfer characteristic of an op amp due to practical limitations: (*a*) ideal, (*b*) finite voltage gain, (*c*) double-valued g_m, (*d*) crossover distortion, (*e*) nonzero input-offset voltage, (*f*) saturation and clipping, (*g*) hysteresis effect.

characteristics of an ideal operational amplifier. Since an ideal device has almost infinite open-loop gain, the plot of the output voltage v_{out} versus the differential input voltage v_D is a straight, nearly vertical line. Thus for an ideal op amp, the noninverting and inverting input voltages are constrained to be equal.

A finite voltage gain affects the ideal op amp performance by causing the transfer characteristic to rotate clockwise such that the slope is no longer infinite (see Figure 5.33b). This slope can be related to the voltage gain of the op amp model of Figure 5.32 through

$$\frac{v_{out}}{v_D} = A_v = g_m R_L \qquad (5.46)$$

We are assuming low-frequency operation where the reactance of C_{comp} is very large compared with R_L. We also assume that the input signal v_D is small enough that i_C is linearly related to v_D through the transconductance g_m.

If the transconductance is not constant but varies with the polarity of the applied signal v_D, the transfer characteristic of the op amp will be double-valued as in Figure 5.33c. Here the gain of the op amp is greater in the negative direction than in the positive direction.

Cross-over distortion occurring in the output stage can cause discontinuities in the transfer characteristic (see Figure 5.33d). This type of distortion is due to the bias applied to the output transistors, which causes either class AB or class B operation. Thus the response of the output stage in the neighborhood of the origin is nonlinear.

The effect of a nonzero offset voltage is illustrated in Figure 5.33e. The transfer characteristic no longer passes through the origin but is translated by a voltage equal to the input-offset voltage V_{IO}.

Device limiting is apparent in Figure 5.33f, where one or more of the amplifier's stages become saturated when the output voltage approaches the power-supply levels. Once full saturation occurs, the output voltage becomes clipped at a limiting value determined mainly by the positive and negative power-supply voltages.

Figure 5.33g shows the hysteresis effect that can occur in an amplifier when the trace and retrace cycles used to produce the transfer characteristic do not coincide. The most probable cause of the hysteresis is localized heating of transistor pairs in the op amp. When the applied voltage v_D is increasing, one transistor of the pair begins to conduct appreciable current and generates heat. When v_D decreases, this transistor conducts less current and cools, while the other transistor in the pair generates heat. A thermal gradient is set up across the chip which causes a shift in the input-offset voltage. The effects of these practical limitations radically alter the ideal transfer characteristics of an operational amplifier. The various effects we have just considered are called static properties because they are basically dc limitations that alter the ideal operation of operational amplifiers.

Dynamic Response

Dynamic limitations to ideal operational amplifiers are generally described by the bandwidth, the transient response, and the slew-rate limitations of the op amp. If an "ideal" op amp should have infinite frequency response, why compensate the op amp to limit its response? The reason is that the "ideal" amplifier must also have zero phase shift (or at least less than $-180°$) over that infinite frequency, and this condition is obviously unattainable. The goal of frequency compensation is to ensure that the loop gain is less than unity at a frequency lower than that at which the phase shift reaches 180°.

A somewhat more practical goal for an op amp response is a -6 dB/octave slope over most of the frequency region of greater than unity loop gain.

Figure 5.34 shows why compensation is necessary. The frequency of the first pole, f_1, is set by the internal compensation. If the phase shift continued to remain flat at $-90°$ far beyond the frequency for unity loop gain, f_2, there would be no problem. At some high frequency in any amplifier, however, parasitic capacitance and transistor response limitations cause the response slope to exceed -6 dB/octave, and the phase shift is made to pass through $-180°$.

Note that "loop gain" is used on the plots, rather than open-loop gain. This is the open-loop response through the amplifier plus the feedback network, and it immediately indicates whether the circuit will be stable when the loop is closed. In Figure 5.34a the phase shift is $-180°$ at a frequency lower than that at which the gain reaches 0 dB (f_2); thus the circuit would oscillate when the loop is closed. In Figure 5.34b additional compensation has been added to the amplifier, moving f_1 to a lower frequency but having little or no effect on the excess phase shift at higher frequencies. The result is that f_2 moves to a frequency lower than that where the phase shift is $-180°$, and we can be confident that the circuit is now stable.

Referring to Figure 5.32, it is evident that the low-frequency corner is determined by

$$f_1 = \frac{1}{2\pi R_L C_{comp}} \tag{5.47}$$

Since R_L is a very high value determined by output and input impedances

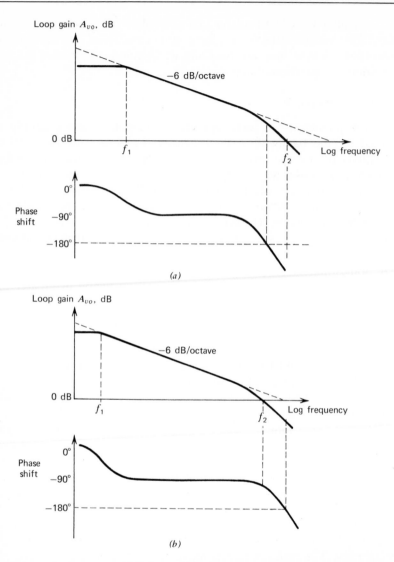

Figure 5.34 Bode plots. (*a*) Insufficient compensation. (*b*) Adequate compensation.

at the compensation point, we might expect f_1 to change considerably with changes in temperature and it does. However these changes in f_1 have a very minor effect on the op amp closed-loop stability. As illustrated in Figure 5.35, if f_1 lowers to f_1' or raises to f_1'' the amplifier dc gain changes inversely by the same amount, leaving the bulk of the response curve undisturbed. This can be shown by computing the position

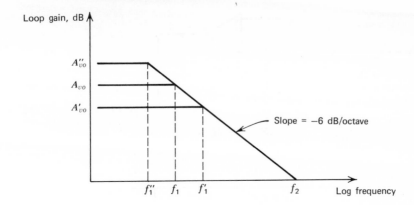

Figure 5.35   Illustration of how f_2 is independent of changes in f_1 with temperature.

of the -6 dB/octave line as given by the gain-bandwidth product from the op amp model. The gain–bandwidth product at f_1 is

$$GBW = A_{vo} \times f_1 = g_m R_L \times \frac{1}{2\pi R_L C_{\text{comp}}}$$

$$= \frac{g_m}{2\pi C_{\text{comp}}}$$

(5.48)

which shows that GBW is independent of R_L but does depend on g_m, which may be designed to be fairly stable over temperature.

The transient response comprises the rise time and overshoot of the device under given input-signal and loading conditions. These conditions must be specified such that the op amp remains in its linear region without becoming slew-rate limited because of saturation effects.

Op amps used in stable feedback networks generally have well-behaved transfer functions, exhibiting an output voltage that closely approximates the solution of a first-, second-, or third-order, linear differential equation. Figure 5.36 shows a feedback configuration and typical wave shapes for determining the transient-response characteristics of an op amp. In the test circuit (Figure 5.36a), the ideal closed-loop voltage gain is set by the two feedback resistors according to

$$G_+ = \frac{v_{\text{out}}}{v_{\text{in}}} = 1 + \frac{R_2}{R_1}$$

(5.49)

Figure 5.36b shows normalized input and output wave shapes produced by the op amp in response to a pulse input. Delay times associated with the internal response of the amplifier immediately after changes in the

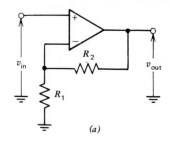

(a)

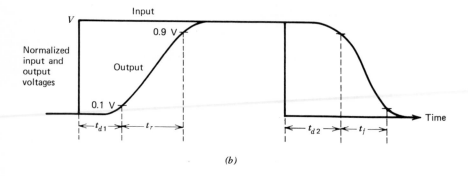

(b)

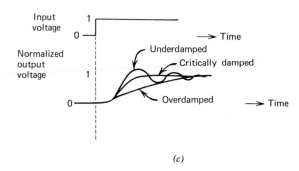

(c)

Figure 5.36 Transient response of an op amp. (a) Test circuit. (b) Identifications of key time intervals. (c) Typical output voltages.

input signal are denoted as t_{d1} and t_{d2}. These delays are due to the finite time required for the input signal to propagate through the various stages of the op amp. The rise and fall times, t_r and t_f. respectively, are measured from the 10 and 90% points of the output-voltage wave shape and are indicative of the time necessary for the amplifier to respond fully to the input signal. As a good first-order approximation, the rise and fall times are related to the upper -3 dB frequency of the op amp by

$$t_r = \frac{0.35}{f_{-3\text{dB}}} \qquad (5.50)$$

We now consider the effect of varying the closed-loop voltage gain on the transient response (rise time and overshoot) of a stable op amp connection (see Figure 5.36a). At unity gain where $R_1 = \infty \Omega$ and $R_2 = 0$ (voltage-follower connection), the output overshoots the final value before going into a damped oscillation about the final value (see Figure 5.36c). The minimum rise time with the maximum overshoot occurs for this condition of $G_+ = 1$. As the closed-loop gain is increased, the overshoot decreases and the rise time increases. The critical damping point occurs where there is no overshoot and the response is monotonic. Further increases in the closed-loop gain continue to reduce the rise time, producing an output similar to that of an R-C low-pass filter.

The slew rate of an op amp is the second important criterion associated with the amplifier's dynamic response. Of all the common op amp parameters, slew rate is one of the least understood, particularly in its relationship to other parameters. We first discuss the slewing phenomenon and indicate how it results from fundamental circuit design limitations related to other device parameters.

We can understand the slew-rate limitations of an amplifier by referring to the simple model illustrated in Figure 5.32. Under steady-state conditions the transconductance stage furnishes only a very small amount of current to R_L. Under transient conditions, however, current must also be furnished to charge or discharge C_{comp}.

In Figure 5.37 the output current of the transconductance stage i_C appears as a function of the differential-input voltage v_D. In general, the transconductance $\Delta I_C / \Delta V_D$ is linear over a fairly wide range of v_D; this range runs from about ± 10 mV to ± 2 V in different amplifier types. Beyond this linear region, i_C no longer increases with an increase in v_D because the transconductance stage cannot deliver an unlimited amount of current. In some amplifier types, the maximum available current $I_{\max}$ is different in the positive and negative directions, leading to different slew rates when the amplifier is slewed in different directions. This current-saturation effect should not be confused with output-voltage saturation of the amplifier, since in most amplifiers $I_{\max} R_L \gg V_{\text{supply}}$. In steady-state operation the amplifier output reaches voltage saturation with values of v_D much lower than are necessary to reach $I_{\max}$. It is only during transient conditions, when extra current is required from the transconductance stage, that $I_{\max}$ is approached.

Figure 5.38 illustrates the basic difference between transient response

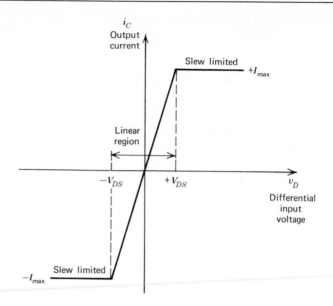

Figure 5.37   Output current versus differential input voltage for transconductance amplifier.

and slewing response. Although the input and output wave forms are fairly similar in appearance, quite different events are taking place within the amplifier. A voltage-follower connection is illustrated for simplicity. In either case, when a step input is applied, the differential input voltage v_D instantaneously is the same value as the input step, since the output has not yet responded. (In an inverting unity-gain connection, only half the input step will appear across the inputs, since the input and feedback resistors act as a divider.) In Figure 5.38b, v_D is less than that required to produce I_{max} (V_{DS}); thus i_C will always be directly proportional to v_D. As the output starts to rise, v_D becomes smaller, making i_C proportionally smaller and reducing the charging rate of C_{comp}. The resulting output wave form for a "model" amplifier is a simple exponential function. In most amplifiers in which the open-loop phase shift at 0 dB loop gain exceeds $-90°$, the actual wave form will exhibit overshoot and a damped oscillation about the final value. However, if the actual frequency-response characteristics are known, the rise time, overshoot, and settling time for the circuit can be predicted by the solution of a linear differential equation. As long as v_D remains below V_{DS}, the output rise time will be constant regardless of the input step amplitude. Therefore, the rate of change of the output voltage, $\Delta V_{out}/\Delta T$, will be directly proportional to the amplitude of the input step.

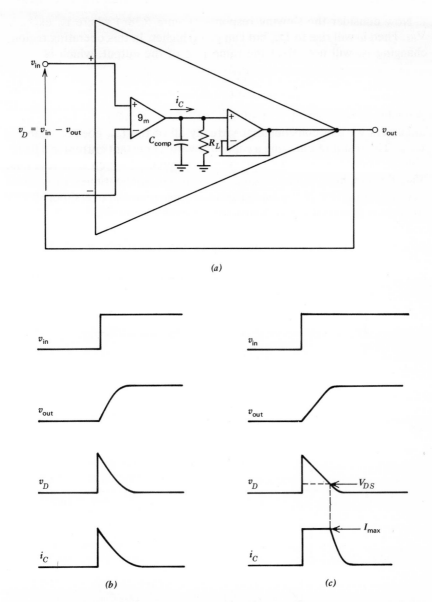

Figure 5.38 Transient response versus slewing response. (*a*) Test circuit. (*b*) Wave shapes for $v_{in} < V_{DS}$. (*c*) Wave shapes for $v_{in} > V_{DS}$.

Now consider the slewing response (Figure 5.38c) where v_D exceeds V_{DS}. Then i_C will rise to I_{max} but can go no higher. In this operating region, changing v_D will not affect the ramp rate of the output, which is

$$\frac{\Delta V_{out}}{\Delta T} = \frac{I_{max}}{C_{comp}} \tag{5.51}$$

Actually, a very small part of I_{max} will flow through R_L, but in most amplifiers we can consider that virtually all of the I_{max} goes to charging C_{comp}. The result of charging a capacitor with a constant current is a linear voltage ramp. When the output rises to the point at which v_D is less than V_{DS}, the amplifier stops slewing, and the output continues to its final value. This portion of the wave form has a rise time and overshoot similar to what is obtained under transient-response conditions.

The critical differential input voltage V_{DS}, seldom indicated on an op amp data sheet, can be computed from other available parameters. To derive this value, consider an op amp connected open-loop with a small step input, as in Figure 5.39. The output would tend to rise as a simple

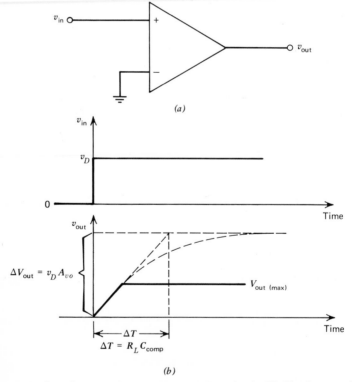

Figure 5.39 Open-loop transient response. (a) Test circuit. (b) Circuit wave shapes.

exponential wave form, having a final value of $v_D \times A_{vo}$ and a time constant of $R_L C_{comp}$. Of course, the amplifier output voltage will probably saturate long before reaching this final value. Thus the visible waveform will approximate a linear ramp with a slope of

$$\frac{\Delta V_{out}}{\Delta T} = \frac{v_D A_{vo}}{R_L C_{comp}} = \frac{v_D g_m R_L}{R_L C_{comp}} = \frac{v_D g_m}{C_{comp}} \qquad (5.52)$$

However, from Equation 5.48 we have

$$\frac{\Delta V_{out}}{\Delta T} = 2\pi v_D GBW \qquad (5.53)$$

This equation is a useful expression for calculating the rise time of an op amp connected as a comparator.

Since slew rate is the maximum value of $\Delta V_{out}/\Delta T$, we can now compute the minimum v_D required to slew the amplifier, V_{DS}.

$$V_{DS} = \frac{SR}{2\pi GBW} \qquad (5.54)$$

From the foregoing, we can draw some conclusions about slewing behavior and clear up some common misconceptions:

1. The amplifier will be slew-rate limited in its response only if v_D exceeds V_{DS}; otherwise the response will be governed by the small-signal frequency-response characteristic for the amplifier in its particular feedback configuration. Putting it another way, if we compute the output rise time from the small-signal characteristics, the amplifier will be slew limited only if the computed

$$SR < \frac{\Delta V_{out}}{\Delta T} \qquad (5.55)$$

2. If C_{comp} remains constant, the slew-rate limit is independent of the feedback configuration, even though the transient response is highly dependent on the closed-loop configuration. In an externally compensated amplifier, we can use smaller values of C_{comp} at higher closed-loop gains, thus raising the slew rate. But since at higher gains we usually have smaller input steps, v_D might not exceed V_{DS}, and the small-signal transient-response characteristic will dominate.

3. Slew rate depends chiefly on I_{max} and C_{comp}. Since I_{max} does not affect bandwidth or other small-signal parameters, there is no fixed relationship between bandwidth and slew rate. Depending on its design, an amplifier may have a wide bandwidth with or without high slew rate. Probably the best design is one in which the time required for the amplifier to slew over its full range approximates its small-signal rise time as a voltage follower.

4. Overshoot characteristics at the end of a slewing wave form are dependent on the circuit closed-loop configuration; thus it is possible to modify these characteristics with external networks, such as Section 5.9 indicates, to optimize settling time.

5. With high-capacitance loads, the output current capability of the amplifier may determine the slew rate:

$$SR = \frac{I_{out(max)}}{C_{load}} \qquad (5.56)$$

As an example of the points just listed, consider the interesting problem of attempting to use an op amp in the open-loop manner to replace a voltage comparator. Suppose we want a moderately fast comparator function: with a 5 mV input step, the output should traverse 5 V in less than 100 ns.

First the necessary slew rate is obvious:

$$SR = \frac{\Delta V_{out}}{\Delta T} = \frac{5 \text{ V}}{100 \times 10^{-9} \text{ s}} = 50 \text{ V}/\mu\text{s} \qquad (5.57)$$

Then the necessary gain–bandwidth product can be computed from Equation 5.53 as

$$GBW = \frac{SR}{2\pi v_D} = \frac{50 \times 10^6}{5 \times 10^{-3} \times 6.28} \text{ Hz} = 1600 \text{ MHz} \qquad (5.58)$$

This equation shows why even very fast op amps do not make particularly fast comparators. Since the comparator is not designed to be stable with negative feedback, it can be designed for maximum gain bandwidth without regard for phase shift. Also, the comparator is usually designed for fast recovery from output saturation conditions. Since little regard for this feature is reflected in most op amp designs, the op amp usually has a longer delay time before its output starts to change.

Power Bandwidth

Since slew rate is the fastest possible rate of change of the op amp's output voltage under large-signal conditions, the slew rate places a limit on the large-signal handling capabilities of high-frequency sine waves. The maximum slope of a sine wave occurs at the zero crossing point and is given by

$$\left(\frac{dv}{dt}\right)_{max} = 2\pi f V_p \qquad (5.59)$$

where f is the frequency and V_p is the peak value of the sine wave.

Substituting the slew rate for $(dv/dt)_{max}$ gives

$$(V_p)_{max} = \frac{SR}{2\pi f} \qquad (5.60)$$

This equation represents the maximum peak-amplitude sine wave the amplifier can provide without distortion at a particular frequency, regardless of closed-loop gain. Like slew rate, this voltage is set by the amplifier's internal saturation levels and is not related to signal bandwidth (i.e., transient response).

Voltage and Current Offsets and Bias Current

The equivalent circuit of Figure 5.40 is useful for studying the effects of offset voltage and current on the performance of the op amp. In

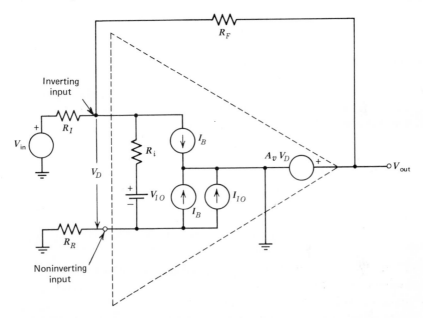

Figure 5.40 Model of op amp for study of voltage and current offsets and bias current. (From *Electronic Integrated Circuits and Systems* by F. Fitchen © 1970 by Litten Educational Publishing, Inc. Reprinted by permission of Van Nostrand Reinhold Company.)

developing this equivalent circuit, we have neglected the amplifier's output impedance and have considered the open-loop voltage gain to be A_v. In our analysis we are concerned only with dc currents and voltages. Hence the signal V_D is the dc portion of the input signal that reaches the internal input resistance R_i. The other three resistors in the figure are all external to the op amp and are used for setting the closed-loop gain and minimizing the effect of the bias current, I_B. The input-offset current and voltage of the op amp are represented by I_{IO} and V_{IO}, respectively.

Fitchen (9) has shown that the output voltage can be expressed as a function of the four quantities V_{in}, V_{IO}, I_B, and I_{IO} according to

$$V_{out} = -\frac{KV_{in}}{D} + \frac{(K+1)V_{IO}}{D} + \frac{[KR_I - (K+1)R_R]I_B}{D} - \frac{(K+1)R_R I_{IO}}{D} \quad (5.61)$$

The quantity K is equal to the ratio of the external feedback resistance R_F to the external input resistance R_I or

$$K = \frac{R_F}{R_I} \quad (5.62)$$

The denominator of Equation 5.61 is given by

$$D = 1 + \frac{K}{A_v}\left[\frac{1}{R_i}\left(R_I + R_R \frac{K+1}{K}\right) + \frac{K+1}{K}\right] \quad (5.63)$$

Equation 5.63 shows that the factor D approaches unity as the open-loop voltage gain A_v approaches infinity. The final three terms on the right-hand side of Equation 5.61 clearly show the effects of the input-offset voltage and bias and offset currents.

The bias-current term I_B can be eliminated by making

$$R_R = R_I \parallel R_F \quad (5.64)$$

When this is done, the expression for the output voltage becomes

$$V_{out} = -\frac{KV_{in}}{D} + \frac{(K+1)V_{IO}}{D} - \frac{(K+1)R_R I_{IO}}{D}$$

$$= -\frac{KV_{in}}{D} + \frac{(K+1)V_{IO}}{D} - \frac{R_F I_{IO}}{D} \quad (5.65)$$

This equation shows that removing the effects of the bias current I_B will not eliminate the effects of the input-offset voltage or input-offset current. Equation 5.65 also reveals that the $K+1$ factor causes the voltage and current offsets to be twice as large for a -1 gain amplifier as that for a voltage follower (or buffer amplifier).

Finally, Equation 5.65 indicates that input voltage and current offsets are multiplied by approximately the same gain factor, K, as the signal voltage V_{in}. Therefore, unless some external adjustment is made to compensate for these input offsets, their effect on the output voltage cannot be distinguished from the input-signal voltage.

External adjustment of the input-offset voltage is frequently accomplished through external connections to the op amp. Such adjustments typically are made using a variable resistor connected between two external op amp terminals. The wiper of the variable resistance is normally connected to one of the power supplies and is adjusted until the output voltage approaches zero volts with both inputs grounded.

Thermal Effects

Zeroing the output voltage of an op amp by adjusting the input-offset voltage is usually done at room temperature of approximately 25°C. However, as we cited in Figure 5.1, the parameters of practical op amps are sensitive to environmental effects such as temperature changes. Thermal effects can produce detectable changes in the op amp's performance parameters. Usually the most significant parameters that drift with temperature are input-offset voltage and current and bias current.

Figure 5.41 gives the variation of the input-offset voltage for no

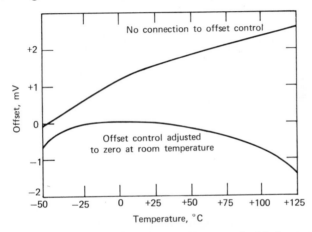

Figure 5.41 Typical offset voltage versus temperature for bipolar op amp.

adjustment to the offset control and for zeroing the offset at room temperature. The shapes of these curves cannot be regarded as "typical," since offset can be of either polarity. In most op amp designs, adjustment of the external offset control also changes the temperature coefficient of the offset voltage; zeroing the offset voltage at room temperature usually (but not always) results in improved temperature stability. However, if the offset control is also used to compensate for offset-current effects or externally generated offsets, poorer temperature stability may result. Better performance might be achieved by summing the voltage from a potentiometer connected between the power lines at one of the amplifier inputs. If selected fixed resistors are used in place of the adjustment potentiometer, it is usually best to simulate the potentiometer with two resistors totaling to the recommended potentiometer resistance.

Typical variations of the bias and offset currents are presented in Figure 5.42. The majority of op amps with bipolar input stages have

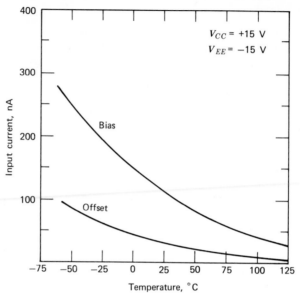

Figure 5.42　Offset bias-current variations with temperature for typical bipolar operational amplifier.

negative temperature coefficients for both bias and offset currents. Bias current is a maximum at the low-temperature extreme and decreases with increasing temperature. The offset-current curve shows the same trend as the bias current. The values of offset current over the temperature range are approximately one-third of those for the bias current at a corresponding temperature for this particular device.

The variation of bias current with temperature and power-supply changes of a typical op amp is depicted in Figure 5.43a. Figure 5.43b indicates that op amp supply current requirements are sometimes increased with increasing temperature.

Open-loop voltage gain of an op amp also is sensitive to temperature, as Figure 5.44 reveals. The gain for this device at 125°C is reduced by approximately 10 dB from that achieved at room temperature. Temperatures below 25°C have less of an effect on gain than for the higher values.

The ability of an op amp to dissipate power in the form of heat energy is

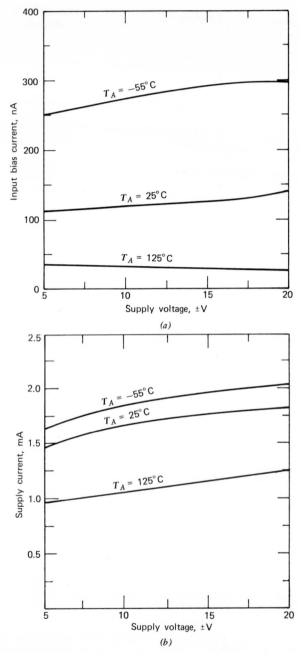

Figure 5.43 Variations in (a) bias current and (b) supply-current, due to temperature and supply-voltage change.

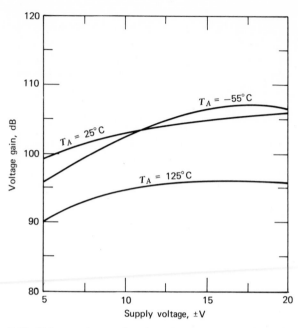

Figure 5.44 Voltage gain as a function of temperature and supply voltage.

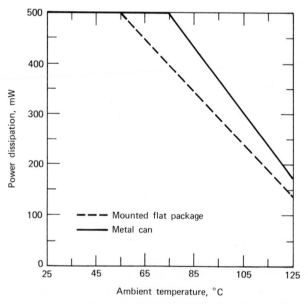

Figure 5.45 Maximum power dissipation versus temperature.

seriously affected at higher temperatures, as we see in Figure 5.45. The device's maximum power dissipation must be derated above a certain temperature, depending on the packaging configuration. The specific amount of derating also depends on packaging. The maximum power dissipation at any temperature can be found according to

$$P_D(T) = \begin{cases} P_{D(max)} & \text{for} \quad T < T_0 \\ P_{D(max)}[1 - \theta(T - T_0)] & \text{for} \quad T \geqslant T_0 \end{cases} \tag{5.66}$$

where θ is the thermal conductance and T_0 is the derating temperature. For the specific device characteristics shown in Figure 5.45, Equation 5.66 can be applied as

Flat Pack	Metal Can
$T_0 = 55°C$	$T_0 = 75°C$
$\theta = 5.21$ mW/°C	$\theta = 6.5$ mW/°C
$P_{D(max)} = 500$ mW	$P_{D(max)} = 500$ mW

Noise

Signal-to-noise ratio is a very important consideration in the design of communications and control systems. The resolution and accuracy of these systems is highly dependent on utilizing designs and devices that achieve low noise figures and low signal-to-noise ratios.

Two dominant types of noise are generated within operational amplifiers. One type is thermal noise, which is broad-band white noise whose magnitude is reduced by reducing the amplifier's bandwidth. Resistors in analog ICs also contribute thermal noise, and it is desirable to keep all resistance values to a minimum because the broad-band thermal noise of a resistor is proportional to the square root of the resistance.

The other type of dominant noise in an op amp is $1/f$ noise, which behaves as the name implies, being inversely proportional to frequency. As a rule, $1/f$ noise is negligible above 10 kHz.

The input stage is the prime noise contributor within an op amp if the gain of the first stage is high. Usually the noise contributed by later stages is small compared with the amplified noise of the first stage.

Figure 5.46 displays typical op amp noise levels referred to the input versus op amp bandwidth for 0 and 10 kΩ values of source resistance. For comparative purposes, the thermal noise generated in a 10 kΩ resistor at 25°C is also shown. This thermal noise is calculated according to

$$\sqrt{E_n^2} = \sqrt{4kTRB_n} \tag{5.67}$$

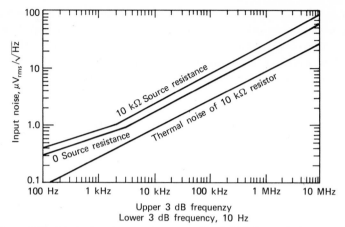

Figure 5.46 Equivalent input noise versus bandwidth for typical op amp.

where $\sqrt{E_n^2}$ is the rms thermal noise, k is Boltzmann's constant $(1.38 \times 10^{-23} \text{ J/}^{\circ}\text{C})$, T is temperature in degrees Kelvin, R is the resitance value in ohms, and B_n is the effective bandwidth in hertz.

Experimental noise curves must be obtained by using a true rms voltmeter having a low -3 dB frequency as small as possible. The bandwidth of the op amp is easily controlled by using various compensation capacitors at the external bandwidth control points. The bandwidth of internally compensated op amps can be set by using various closed-loop gains.

A special note must be inserted concerning bandwidth determination. The effective noise bandwidth is the bandwidth of a filter having a rectangular frequency response (22). Such a response is not easily achieved in practice. The effective bandwidth of an amplifier can be determined by integrating the area under the frequency-response curve. This area is then divided by the maximum amplifier gain.

Most op amps have a single corner frequency f_h, with a 20 dB/decade rolloff. The effective noise bandwidth of the amplifier B_n is related to f_h as

$$B_n = 1.57 f_h \qquad (5.68)$$

The bandwidth given in Figure 5.46 is the -3 dB bandwidth f_h, not the effective noise bandwidth B_n. The zero-source-resistance curve shows the equivalent input noise generated within the op amp. The $10 \text{ k}\Omega$ source-resistance curve contains the sum of the internal equivalent noise, the broad-band thermal noise of a $10 \text{ k}\Omega$ resistor, and the noise voltage developed across the $10 \text{ k}\Omega$ resistor because of noise current. The predominant noise at 100 Hz is the $1/f$ noise. Generally between 1 and 10 kHz, the thermal noise becomes more important.

Using a fairly straightforward graphical approach, broad-band and spot noise curves can be used to establish an overall noise "figure of merit" for an amplifier configuration. This noise figure can then be used to predict the amplifier's total output noise and as a basis for comparing operational amplifier devices. The noise figure is obtained through a combination of the voltage, current, and resistor noise curves plotted in Figure 5.47. The

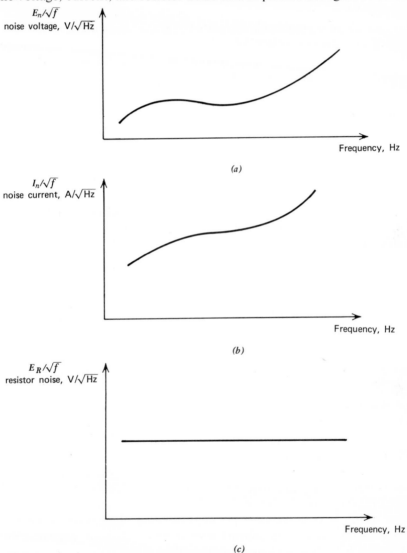

Figure 5.47 General noise curves for determination of equivalent noise figure.

ordinate values of each of the three curves are first squared. All three curves will have units of V^2/Hz after the noise-current curve is multiplied by the square of the effective source resistance. Next, the three squared curves are summed to produce a composite noise curve given by

$$\frac{E_T^{\;2}}{f} = \frac{E_n^{\;2}}{f} + \frac{I_n^{\;2} R_s^{\;2}}{f} + \frac{E_R^{\;2}}{f} \qquad (5.69)$$

This total curve is graphically integrated over the bandwidth of interest to produce an effective noise voltage, $E_T^{\;2}$. The equivalent noise voltage (ENV) for the amplifier then is simply

$$\text{ENV} = \sqrt{E_T^{\;2}} = E_T \; \text{V} \qquad (5.70)$$

Another type of noise found in integrated circuits is "popcorn" noise or burst noise, so named because when the noise is amplified and used to drive a loudspeaker, a sound like corn popping results. Popcorn noise is difficult to define thoroughly. However, it is generally recognized that this noise is more pronounced at low frequencies, low temperatures, and high impedance levels. Furthermore, it is known that popcorn noise is more noticeable in circuits contaminated with certain impurities during device fabrication, leading to the strong suspicion that the noise is a surface-related phenomenon.

5.8 MEASUREMENT OF OPERATIONAL AMPLIFIER PARAMETERS

Most of the static parameters of op amps can be measured quite simply by using the test procedures described in this section. We discuss the measurement of the following parameters:

1. Input-offset voltage.
2. Input-bias current.
3. Input-offset current.
4. Open-loop voltage gain.
5. Common-mode rejection ratio.
6. Power-supply rejection ratio.
7. Output voltage and current capability.
8. Power dissipation.

Many of the procedures cited for determination of op amp parameters apply equally well to measuring similar parameters of voltage comparators (e.g., bias and offset currents, offset voltage, open-loop voltage gain).

The offset voltage, bias and offset currents of the amplifier under test (AUT) can be measured using the circuit presented in Figure 5.48. After setting the power-supply voltages $+V_{CC}$ and $-V_{EE}$ to the desired values,

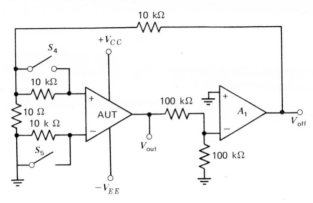

Figure 5.48 Test circuit for measurement of offset voltage, bias current, and offset current.

switches S_4 and S_5 are closed and the voltage V_{off} is measured. The offset voltage of the AUT is equal to $V_{off} \times 10^{-3}$. This simple relationship results because of the feedback action of the circuit and the large voltage gain supplied by the auxiliary amplifier A_1 and the AUT. The V_{out} voltage will be driven to the ground reference voltage by V_{off} to counteract the offset voltage developed across the 10 Ω resistor. Because of voltage division between the 10 kΩ feedback resistor and the 10 Ω resistor, V_{off} adjusts to a value of 10^3 times as large as the actual offset voltage of the AUT to drive V_{out} to zero.

The bias and offset currents are determined by making several measurements of V_{off} under different S_4 and S_5 switch conditions. The voltage V_{off} is measured as before for the offset-voltage determination and recorded as $V_{off\ 1}$. Switch S_4 is then opened and V_{off} is measured as $V_{off\ 2}$. The bias current for the noninverting input is found from

$$I_{B+} = \frac{V_{off\ 2} - V_{off\ 1}}{10^7} \qquad (5.71)$$

where I_{B+} will have units of amperes for V_{off} measured in volts. The factor of 10^7 is a result of the 10 kΩ feedback resistor together with current splitting between the 10 Ω and 10 kΩ input resistors. The bias current for the inverting input, I_{B-}, is found by closing S_4, opening S_5, and remeasuring V_{off} ($V_{off\ 3}$). We calculate I_{B-} from the following relation:

$$I_{B-} = \frac{V_{off\ 3} - V_{off\ 1}}{10^7} \qquad (5.72)$$

The bias current is equal to the average of I_{B+} and I_{B-}, or

$$I_B = \tfrac{1}{2}(I_{B+} + I_{B-}) \tag{5.73}$$

The offset-bias current is the absolute value of the difference of I_{B+} and I_{B-}, or

$$I_{IO} = |I_{B+} - I_{B-}| \tag{5.74}$$

Determination of the open-loop voltage gain of an op amp is a delicate measurement requiring good experimental techniques. The circuit in Figure 5.49 is a useful test configuration; it reduces many of the dominant

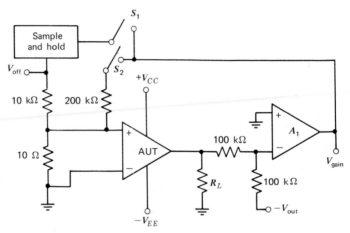

Figure 5.49   Test circuit for measuring open-loop voltage gain.

disturbances that typically degrade this sensitive measurement. Data from an open-loop voltage-gain measurement are meaningful only if taken on a circuit that has been stabilized against input-offset voltage and currents. Circuit stability is first achieved by keeping the feedback loop closed and measuring the input-offset voltage that keeps the AUT in its linear region. The specific measurement procedure is as follows.

The desired test conditions are established by setting the $+V_{CC}$ and $-V_{EE}$ supply voltages to the specified values. Also the prescribed load is applied as R_L. Initially switch S_1 is closed, S_2 opened, and $-V_{out}$ grounded; under these conditions, the sample-and-hold circuit will null the offset voltage. After the circuit stabilizes, S_2 is closed and S_1 opened, with the sample-and-hold now maintaining a voltage sufficient to null the input-offset voltage. Next $-V_{out}$ is set to one limit of the desired output voltage level, $-V_{out\,1}$, and the output voltage of A_1 is measured as $V_{gain\,1}$. Then $-V_{out}$ is readjusted to the other desired limit, $-V_{out\,2}$, and $V_{gain\,2}$ is measured. The open-loop voltage gain of the AUT is found from

$$A_v = \frac{V_{\text{out }1} - V_{\text{out }2}}{V_{\text{gain }1} - V_{\text{gain }2}} \times 20,000 \qquad (5.75)$$

The 20,000 factor is the result of voltage division between the 200 kΩ and 10 Ω feedback resistors.

The $-V_{\text{out}}$ could first be set to zero and then to -10 V. Applying the test procedure and Equation 5.75 will give the voltage gain in the plus direction. If 0 and $+10$ V are used for V_{out}, the gain in the negative direction results. The average gain over both positive and negative excursions can be measured by using output voltages of -10 and $+10$ V.

Several techniques can be employed to measure the common-mode rejection ratio of an op amp, the most straightforward being the utilization of a circuit like that in Figure 5.50. The difference-mode and common-

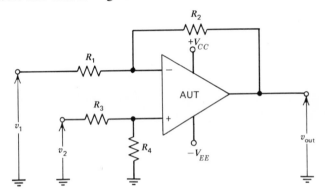

Figure 5.50 Simple circuit for measuring op amp common-mode rejection ratio.

mode gains (see Equations 3.1 and 3.2) are measured under the specific input-signal conditions of $v_1 = v_2$ and $v_1 = -v_2$. The circuit offers simplicity, but accuracy is severely limited by the tolerance of the four resistors. For example, if equal-valued resistors having 1% tolerance were used in this circuit, the accuracy of the test circuit would be approximately 45 dB, much below the typical 60–120 dB range shown in Figure 5.1.

The test circuit in Figure 5.51 overcomes the resistor tolerance limitation for measuring CMRR. The principle on which the operation of this circuit is based is a shifting of the ground-reference voltage by adjustment of power-supply voltages. However, the total power-supply voltage (the sum of absolute values of $+V_{CC}$ and $-V_{EE}$) is always kept constant for any one CMRR measurement.

The CMRR test is performed as follows. Suppose that the total

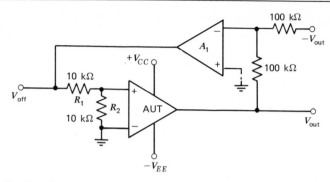

Figure 5.51 Test circuits for measurement of common-mode rejection ratio and power-supply rejection ratio.

power-supply voltage for the test condition is to be 30 V, normally achieved by setting $+V_{CC}=15$ V and $-V_{EE}=-15$ V. If instead we set $+V_{CC}=20$ V and $-V_{EE}=-10$ V, we still have an effective 30 V power supply with a phantom ground at the midway voltage of +5 V. After making these power-supply settings, applying -5 V to $-V_{out}$ will drive V_{out} to the phantom-ground voltage of +5 V because of the feedback configuration of the test circuit. The offset voltage V_{off} is measured as $V_{off\ 1}$. By voltage division, $10^{-3}\ V_{off\ 1}$ is the signal voltage applied to the AUT to drive $V_{out}=+5$ V.

The test procedure is repeated after adjusting the power-supply voltages to $+V_{CC}=+10$ V and $-V_{EE}=-20$ V, which still keeps an effective power-supply voltage of 30 V. The V_{out} is driven to the new phantom ground of -5 V by applying $+5$ V to $-V_{out}$. The offset voltage is again measured, being noted now as $V_{off\ 2}$. The CMRR is calculated from the test data according to

$$(CMRR)_{dB} = 20\ \log_{10}\left[\left(\frac{R_2}{R_1+R_2}\right)\left|\frac{V_{off\ 2}-V_{off\ 1}}{\Delta V_{CM}}\right|\right] \qquad (5.76)$$

where ΔV_{CM} is the change in the common-mode voltage due to shifting the phantom-ground voltages. For our test conditions $\Delta V_{CM}=10$ V and the resistor ratio is 10^{-3}. Thus a 1 V difference in offset voltage is equivalent to a CMRR of -80 dB.

This technique for CMRR features high accuracy without requiring high-tolerance resistors. Furthermore, the gain of the feedback loop enables V_{off} to be measured before it becomes attenuated by R_1 and R_2 and is applied as an input voltage to the AUT. Thus the common problem of having to form the difference between two very small and close-valued quantities is not encountered. Highly accurate measurements of the individual $\pm V_{out}$ quantities also are not required because of the large difference between them. Slight inaccuracies in setting either of the power-supply voltages will be reduced by the power-supply rejection ratio, which is the subject of our next measurement test.

The figure of merit indicating the op amp's ability to compensate for changes in power supply voltages is called the power-supply rejection ratio (PSRR). The test circuit of Figure 5.51 used for CMRR determination can again be employed to measure PSRR. The PSRR test differs only in one respect from the CMRR test in that no longer is the total power-supply voltage maintained constant. For example, suppose nominal $+V_{CC}$ and $-V_{EE}$ power-supply voltages of $+15$ V and -15 V are desired, and we wish to test the sensitivity of $+V_{CC}$ to ± 5 V variations. After grounding $-V_{out}$ and setting $-V_{EE}$ to -15 V, we adjust $+V_{CC}$ to the upper limit of $+20$ V; V_{off} is measured as before and identified as $V_{off\,1}$. Next $+V_{CC}$ is set to $+10$ V and V_{off} remeasured as $V_{off\,2}$. The PSRR for $+V_{CC}$ is then calculated as

$$(PSRR)_{dB} = 20 \log_{10}\left[\left(\frac{R_2}{R_1+R_2}\right)\left|\frac{V_{off\,1}-V_{off\,2}}{+V_{CC}}\right|\right] \qquad (5.77)$$

where $+V_{CC}$ is the nominal $+15$ V.

The PSRR for the $-V_{EE}$ supply is determined in a similar manner by establishing $+15$ V on $+V_{CC}$ and making V_{off} measurements for -20 V and -10 V settings of the $-V_{EE}$ supply.

The output voltage and current capabilities of the op amp can be tested using the test circuit presented in Figure 5.52. A load resistor R_L is connected to the output of the AUT. The value of R_L is chosen to yield an

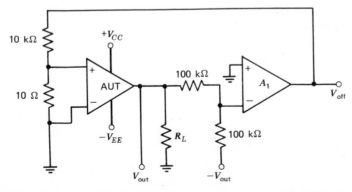

Figure 5.52 Test circuits for measuring output voltage and current capabilities and power dissipation.

output current that is the minimum acceptable output current at the desired output voltage. The AUT is programmed to produce a voltage greater than the desired output level by applying a voltage of equal magnitude but opposite polarity to the desired output voltage at $-V_{out}$. The AUT produces V_{out}, which is measured to see whether it reaches the desired output voltage. This test is performed for both conditions of driving the output voltage—positive and negative.

Power dissipation within the op amp is found by driving V_{out} to zero by grounding $-V_{out}$ and measuring the current supplied by one of the power supplies. Power dissipation in the op amp is the product of this current and the total power-supply voltage.

Figure 5.53 is a complete schematic useful for measuring all op amp

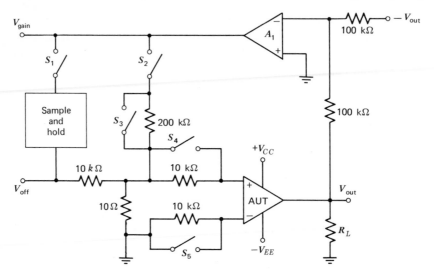

Figure 5.53 Simplified schematic of the complete dc test circuit for operational amplifiers.

parameters discussed in this section. The open and closed positions of the five switches can be arranged in various combinations to produce any of the test circuits cited. Switch positions can be programmed easily under computer control with the circuit of Figure 5.53 for generating an automated test fixture.

5.9 AC ANALYSIS OF OPERATIONAL AMPLIFIER CIRCUITS

One of the more frustrating aspects of electronics design is the occasional tendency of the completed circuit to oscillate, even when the designer did not intend to build an oscillator. This problem usually can be avoided by analysis of the gain and phase-shift characteristics at various frequencies of both the op amp and the external-feedback network. A number of mathematical and graphical procedures have been developed for stability analysis of feedback systems. Most of these techniques require mathematical expressions for the amplifier and network transfer functions. The graphical approach illustrated here requires only readily obtainable curves for the components. The steps are as follows:

1. Obtain the graphs of open-loop gain (in decibels) and phase shift versus frequency for the op amp being used. For externally compensated amplifiers, these curves should be for the amplifier with the proposed compensation network added. If the amplifier data sheet does not include these curves, they generally can be furnished by the manufacturer. The curves could also be physically measured using a vector voltmeter or network analyzer. Figure 5.54 gives typical curves for a fairly wide-band amplifier.

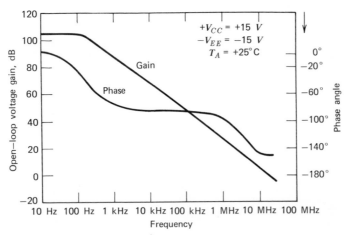

Figure 5.54 Open-loop frequency and phase response of typical op amp.

2. Redraw the feedback network in Figure 5.55a, making the node at the op amp's output the input. As Figure 5.55b shows, the node formerly connected to the op amp inverting input now has become the output terminal of the feedback network. The circuit input is grounded through the source impedance. The op amp's input resistance and input capacitance should be included as part of the network. Draw the gain and phase-angle versus frequency curves for this network alone. For typical

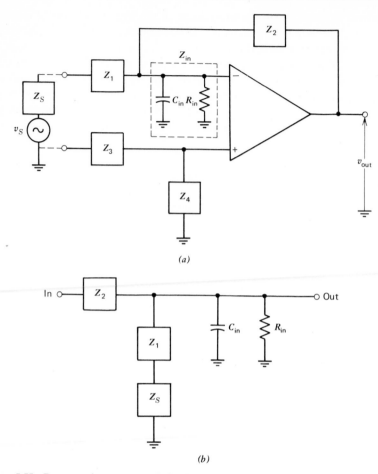

Figure 5.55 Restructuring an op amp's feedback network connection. (*a*) General op amp network. (*b*) Feedback network equivalent circuit.

monolithic op amps, R_{in} is generally 10 to 1000 MΩ and is usually not significant. Usually C_{in} is 2 to 10 pF, which may create a significant phase shift in wide-band systems or when high-valued network resistors are used. Figure 5.56 shows the gain and phase-angle curves for a unity-gain inverting-amplifier connection.

3. Combine the two curves obtained in steps 1 and 2 by algebraically adding the decibel gains and adding the phase angles to obtain a third pair of curves which represent the combined open-loop characteristics from the amplifier input to the feedback point. Observe the phase angle at the frequency where the gain is 0 dB: (*a*) if the phase angle at the 0 dB

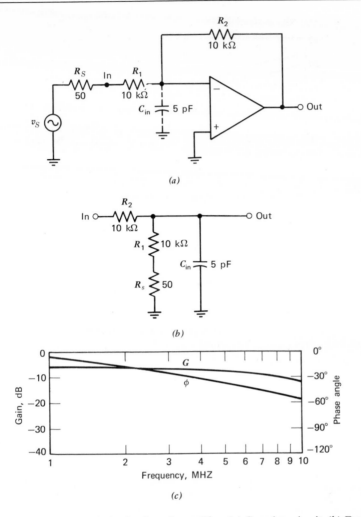

Figure 5.56 Graphical analysis of unity-gain amplifier. (*a*) Complete circuit. (*b*) Feedback equivalent circuit. (*c*) Frequency response of the equivalent circuit.

frequency is $-180°$ or greater, the closed-loop system will oscillate; (*b*) if the phase angle at the 0 dB frequency is less than 180°, the system will be stable.

Figure 5.57 illustrates the combination of the curves in Figures 5.54 and 5.56c. The gain reaches 0 dB at about 6 MHz, but the phase angle is about $-185°$ at this frequency; thus the closed-loop system will probably oscillate. The main problem source is the input capacitance, which creates

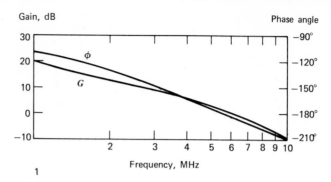

Figure 5.57 Combined curves of Figures 5.54 and 5.56c.

sufficient phase shift in the feedback network to make the system unstable. The solution is to add a small capacitor in parallel with the feedback resistor (see Figure 5.58); this component cancels the phase shift produced by C_{in} and produces a stable system.

In voltage-amplifier circuits, the closed-loop bandwidth is approximately the frequency where the composite gain curve has 0 dB. The open-loop phase angle at this frequency determines the peaking of the closed-loop frequency response, as well as the amount of overshoot under pulsed conditions. At $-90°$ phase shift, the system will be critically damped; near $-180°$, peaking and overshoot will be high.

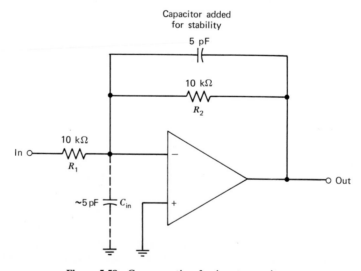

Figure 5.58 Compensation for input capacitance.

Frequency Compensation of Operational Amplifiers

Some monolithic op amps are *uncompensated.* That is, one or more external capacitors and resistors must be connected to the op amp, in addition to the normal feedback components, to ensure stability of the circuit in certain closed-loop configurations.

Internally compensated op amps may be used in nearly any closed-loop configuration without stability problems, since the required compensation components are built into the IC.

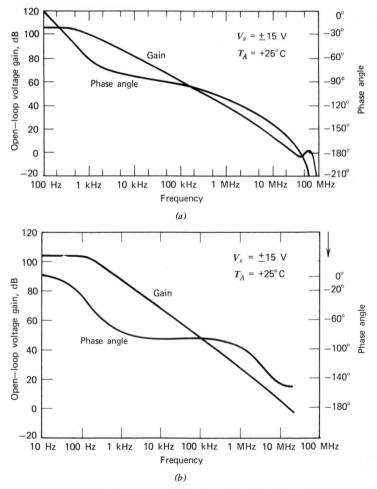

Figure 5.59 Open-loop gain and phase angle versus frequency for (*a*) uncompensated op amp and (*b*) internally compensated op amp.

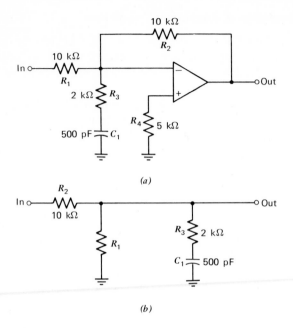

(a)

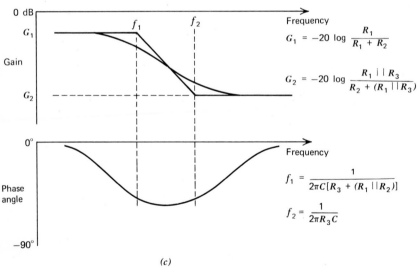

$G_1 = -20 \log \dfrac{R_1}{R_1 + R_2}$

$G_2 = -20 \log \dfrac{R_1 \mid\mid R_3}{R_2 + (R_1 \mid\mid R_3)}$

$f_1 = \dfrac{1}{2\pi C[R_3 + (R_1 \mid\mid R_2)]}$

$f_2 = \dfrac{1}{2\pi R_3 C}$

(c)

Figure 5.60 Input compensation for inverting amplifier. (*a*) Circuit diagram. (*b*) Feedback network equivalent circuit. (*c*) Frequency response of the feedback network.

214

Obviously, the internally compensated type of amplifier is preferable from the standpoints of ease of use and the requirement for fewer external components. However, the uncompensated amplifier is much more versatile in obtaining the best possible ac performance in a given gain configuration.

Figure 5.59a illustrates the open-loop gain and phase angle for a wide-band, uncompensated amplifier; Figure 5.59b shows similar curves for the same basic amplifier design with internal compensation. The uncompensated version would not be stable as a voltage follower, since its gain peaks above 0 dB at frequencies where the phase angle is greater than $-180°$. If a 20 pF capacitor is connected from the bandwidth control pin to ground, the response of the amplifier becomes identical to the compensated amplifier, and it will be stable as a voltage follower.

The uncompensated amplifier can be connected for higher closed-loop gains without compensation. For example, at a closed-loop gain of 10 (+20 dB), the uncompensated amplifier has an available bandwidth of about 7 MHz, versus about 2 MHz for the internally compensated version. Also, the uncompensated amplifier will have a slew rate of about 35 V/μs, compared with approximately 6 V/μs for the compensated version.

A useful compensation scheme for the inverting amplifier connection appears in Figure 5.60a. The advantages of this connection over more conventional compensation schemes are as follows: although bandwidth is reduced, slew rate is not reduced; and high open-loop gain is available at low and medium frequencies.

5.10 CONCLUSIONS

Operational amplifiers were treated in this chapter from two viewpoints. First, the op amp device itself was examined and its performance parameters identified and defined. Second, specific applications of op amps were considered and the functional operation of specific circuits explained. Feedback principles applied to op amps were presented to develop circuit analysis and design capabilities. Techniques for improving circuit performance toward the theoretical optimum were developed and illustrated through practical design examples.

BIBLIOGRAPHY

1. Ahmed, H., and P. J. Spreadbury, *Electronics for Engineers, An Introduction*, Cambridge University Press, London, 1973.

2. Angelo, E. James, Jr., *Electronics: BJTs, FETs, and Microcircuits*, McGraw-Hill, New York, 1969.

3. Barna, Arpad, *Operational Amplifiers*, Wiley, New York, 1971.

4. Belove, Charles, Harry Schachter, and Donald L. Schilling, *Digital and Analog Systems, Circuits, and Devices: An Introduction*, McGraw-Hill, New York, 1973.

5. Cheng, David K., *Analysis of Linear Systems*, Addison-Wesley, Reading, Mass., 1961.

6. Chirlian, Paul M., *Integrated and Active Network Analysis and Synthesis*, Prentice-Hall, Englewood Cliffs, N.J., 1967.

7. D'Azzo, J. J., and C. H. Houpis, *Feedback Control System Analysis and Synthesis*, McGraw-Hill, New York, 1966.

8. Deboo, Gordon J., and Clifford N. Burros, *Integrated Circuits and Semiconductor Devices: Theory and Application*, McGraw-Hill, New York, 1971.

9. Fitchen, Franklin C., *Electronic Integrated Circuits and Systems*, Van Nostrand-Reinhold, New York, 1970.

10. Gardner, Murry F., and John L. Barnes, *Transients in Linear Systems*, Wiley, New York, 1963.

11. Graeme, Jerald G., Gene E. Tobey, and Lawrence P. Huelsman, *Operational Amplifiers: Design and Applications*, McGraw-Hill, New York, 1971.

12. Graeme, Jerald G., *Applications of Operational Amplifiers: Third Generation Techniques*, McGraw-Hill, New York, 1973.

13. Gray, Paul E., and Campell L. Searle, *Electronic Principles: Physics, Models, and Circuits*, Wiley, New York, 1969.

14. Grebene, Alan B., *Analog Integrated Circuit Design*, Van Nostrand-Reinhold, New York, 1972.

15. Hilburn, John L., and David E. Johnson, *Manual of Active Filter Design*, McGraw-Hill, New York, 1973.

16. Huelsman, L. P., *Active Filters: Lumped, Distributed, Integrated, Digital and Parametrics*, McGraw-Hill, New York, 1970.

17. Meyer, Charles S., David K. Lynn, and Douglas J. Hamilton, *Analysis and Design of Integrated Circuits*, McGraw-Hill, New York, 1968.

18. Melen, Roger, and Harry Garland, *Understanding IC Operational Amplifiers*, Sams, Indianapolis, 1971

19. Millman, Jacob, and Christos C. Halkias, *Electronics Devices and Circuits*, McGraw-Hill, New York, 1967.

20. Millman, J., and C. C. Halkias, *Integrated Electronics: Analog and Digital Circuits and Systems*, McGraw-Hill, New York, 1972.

21. Millman, J., and Herbert Taub, *Pulse, Digital, and Switching Waveforms*, McGraw-Hill, New York, 1965.

22. Motchenbaucher, C. D., and F. C. Fitchen, *Low Noise Electronic Design*, Wiley, New York, 1973.

23. Motorola Semiconductor Products Inc., "Linear Integrated Circuits Data Book," Motorola, Inc., December 1972.

24. Pierce, J. F., and T. J. Paulus, *Applied Electronics*, Merrill, Columbus, Ohio, 1972.

25. Pierce, J. F., *Semiconductor Junction Devices*, Merrill, Columbus, Ohio, 1967.

26. Pierce, J. F., *Transistor Circuit Theory and Design*, Merrill, Columbus, Ohio, 1963.

27. RCA Inc., "Linear Integrated Circuits," RCA Corp., Technical Series IC-42, 1970.

28. RCA Inc., "Linear Integrated Circuits and MOS Devices," RCA Corp., No. SSD-202, 1972.

29. Ryder, John D., *Electronic Fundamentals and Applications*, Prentice-Hall, Englewood Cliffs, N.J., 1970.

30. Schilling, Donald L., and Charles Belove, *Electronic Circuits: Discrete and Integrated*, McGraw-Hill, New York, 1968.

31. Sheingold, Daniel H., *Analog-Digital Conversion Handbook*, Analog Devices Inc., Norwood, Mass., 1972.

32. Stewart, Harry E., *Engineering Electronics*, Allyn & Bacon, Boston, 1969.

33. Strauss, Leonard, *Wave Generation and Shaping*, McGraw-Hill, New York, 1970.

34. Su, Kendell L., *Active Network Synthesis*, McGraw-Hill, New York, 1965.

CHAPTER 6

ADVANCED AMPLIFIERS

The basic types of operational amplifier give the circuit and system designer a very useful "gain block" to incorporate into numerous applications for realizing many diverse functional requirements. Although the general op amp can perform a seemingly endless number of functions, some special applications require op amps possessing rather unique characteristics. Usually the characteristics are achieved by combining additional circuit functions with a basic op amp on the same monolithic chip to produce devices we have arbitrarily called advanced amplifiers. In this chapter we present some of these advanced amplifiers, discuss their principles of operation, and show circuits in which their unique characteristics serve special applications.

6.1 PROGRAMMABLE AMPLIFIERS

The programmable amplifiers, a very versatile class of advanced amplifiers, are so named because their performance characteristics can easily be changed to accommodate total system requirements. The external compensation capacitor added to many op amps for feedback stabilization represents one form of amplifier programming. However, this technique of trading stability at low gains for bandwidth, and vice versa, has become so well established that it can hardly be termed an "advanced" amplifier concept.

Our discussion of programmable amplifiers focuses on amplifiers having special provisions for optimizing device response with respect to minimum degradation of other device parameters. We also consider amplifier "gates" where the signal can be programmed through various amplifier paths to achieve some desired signal processing.

Designs using operational amplifiers frequently require tradeoffs to

218

improve some aspect of circuit performance. For example, if an amplifier must have exceptionally good high-frequency performance, such as bandwidth and slew rate, we should expect the op amp to utilize relatively large internal current sources. The price paid for this good ac performance is a reduced dc performance, since large currents introduce high input-current bias levels, contribute to high input-voltage offsets, and require significantly more standby power. On the other hand, a good dc amplifier will usually have small internal current levels, which then limits the dynamic ac performance.

The general-purpose op amp represents a compromise between the two performance extremes of ac and dc conditions. Many applications cannot be satisfied by using the general-purpose op amp. Therefore, the circuit designer must carefully consider the device best suited for his particular application. Moreover, the inventory of the system designer must contain a wide variety of op amps having the necessary specifications for any number of applications in the system.

The programmable op amp allows the designer to tailor the amplifier's characteristics between the limits of low-power, nanowatt operation with excellent dc input characteristics, to the other extreme of high-frequency performance surpassing 10 MHz bandwidth and 6 V/μs slew rate. An op amp offering this selectable range of performance specifications is the HA-2720 (Figure 6.1). Programmability is achieved through control of

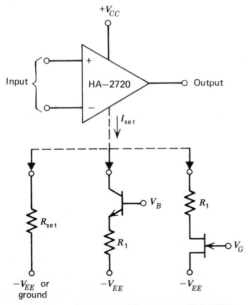

Figure 6.1 Programmable op amp (HA-2720).

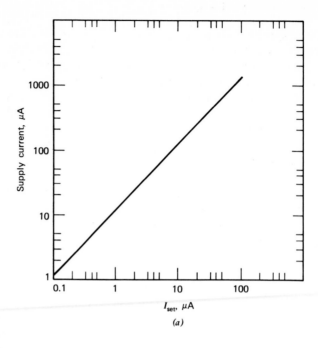

(a)

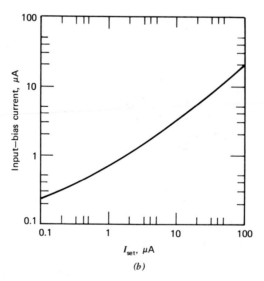

(b)

Figure 6.2 Programmable amplifier parameters as a function of I_{set}: (a) standby supply current, (b) input bias current, (c) slew rate, (d) open-loop gain response, (e) input-noise voltage, (f) input-noise current, (g) minimum noise for a given source resistance, (h) summary of the effect on parameters of increasing I_{set}.

220

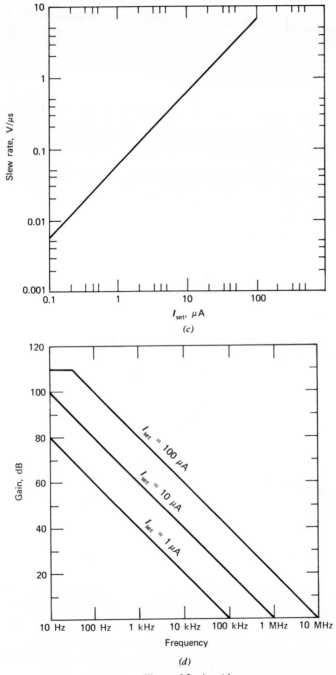

(c)

(d)

Figure 6.2 (cont.)

221

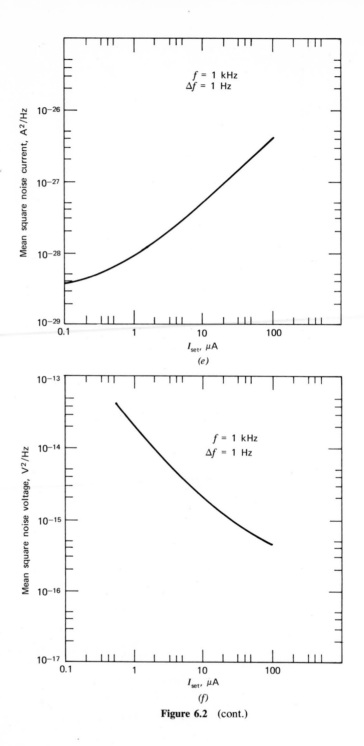

Figure 6.2 (cont.)

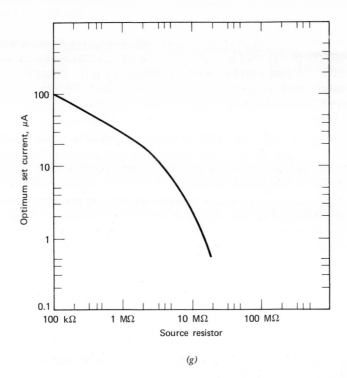

(g)

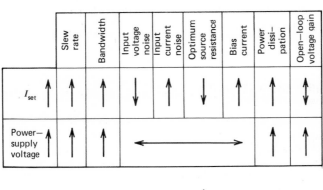

	Slew rate	Bandwidth	Input voltage noise	Input current noise	Optimum source resistance	Bias current	Power dissi- pation	Open–loop voltage gain
I_{set} ↑	↑	↑	↓	↑	↓	↑	↑	↕
Power– supply voltage ↑	↑	↑	←——————————→				↑	↑

←——→ No change   ↑ Increase

↕ Bell–shaped   ↓ Decrease

(h)

Figure 6.2 (cont.)

223

the quiescent current between the device set-current terminal and either the negative power supply or ground. The set current is used to adjust the op amp's internal current sources which, in turn, control the device parameters such as standby current, bias current, slew rate, bandwidth, and input-noise voltage and current. Figure 6.2 shows how the set current controls these op amp parameters. The typical curves illustrate that parameters can be controlled over a three-decade range. Figure 6.2h summarizes the effect produced on the various parameters by increasing the set current. The dynamic programming capability made available by the HA-2720 enables this op amp to serve many special applications, such as being a high-performance amplifier for both ac and dc, or acting as a current-controlled oscillator, a multiplexer, or a sample-and-hold circuit. We now illustrate several of these special applications.

A low-power, dc amplifier configuration set to give $G_- = -10$ appears in Figure 6.3. The set current I_{set} is controlled by R_{set} according to

$$I_{set} = \frac{|V_{CC}| - 2V_{BE}}{R_{set}} \qquad (6.1)$$

where V_{BE} is typically 0.6 V. For $R_{set} = 18$ MΩ, I_{set} becomes 0.1 μA, which produces an exceptionally low input-bias current of 200 pA and a quiescent power dissipation of 6 μW. The closed-loop bandwidth is 1 kHz and the slew rate is 0.006 V/μs.

Changing the R_{set} resistor of Figure 6.3 to 280 kΩ, increasing the power supplies to ±15 V, and applying −15 V instead of ground to R_{set}, transforms the low-power amplifier to a high-performance ac amplifier. The set-current equation is now

$$I_{set} = \frac{|V_{CC}| + |V_{EE}| - 2V_{BE}}{R_{set}} \qquad (6.2)$$

The amplifier now will have a closed-loop bandwidth of 1 MHz, a slew

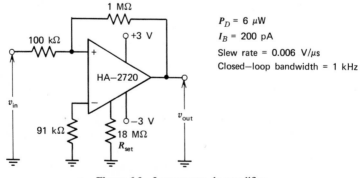

$P_D = 6$ μW
$I_B = 200$ pA
Slew rate = 0.006 V/μs
Closed—loop bandwidth = 1 kHz

Figure 6.3 Low-power dc amplifier.

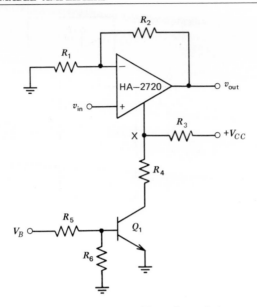

Figure 6.4 Programmable analog switch.

rate of 6 V/μs, and a standby power dissipation of 35 mW. The typical input-bias current is increased to 20 nA, which is still considerably below levels achieved with the general-purpose op amps.

Another programming application of the HA-2720 uses the set current for switching the amplifier "on" and "off" to form an analog switch with gain capability. Such a circuit is presented in Figure 6.4. Applying a positive voltage V_B will saturate transistor Q_1 and establish a path for the set current to turn on the programmable amplifier. The output voltage v_{out} is a function of the feedback network and the input signal according to

$$v_{out} = G_+ v_{in} = \left(1 + \frac{R_2}{R_1}\right) v_{in} \qquad (6.3)$$

The amplifier is switched off by making $V_B < 0$ volts. However, in the off state, Q_1's leakage current may be sufficient to keep the amplifier on. By connecting R_3, a voltage of approximately $+V_{CC}$ is established at point X to prevent the amplifier from turning on.

The programmable amplifier is particularly useful in sample-and-hold circuit applications. The circuit passes (samples) the input signal to the output terminal whenever the control logic is in the logic 1 state. Changing the control logic state puts the amplifier in the blocking mode, where its output maintains (holds) the voltage level existing just prior to changing the control logic state. A circuit that accomplishes the sample-and-hold

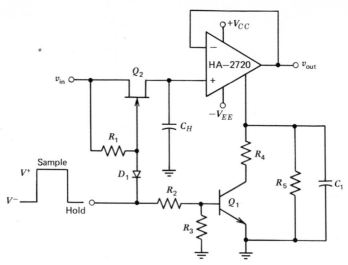

Figure 6.5 Sample-and-hold circuit.

function is diagrammed in Figure 6.5. The sample period exists when the control voltage is positive. The sample signal switches the JFET Q_2 "on," allowing the capacitor C_H to store a portion of the applied input signal. At the same time, the analog input signal is presented to the buffer amplifier, which has been switched on because of the positive sample signal. The sample signal also saturates Q_1, thereby providing a path for the set current. The set current is determined by

$$I_{set} = \frac{V_{CC} - 2V_{BE} - V_{CE(sat)}}{R_4} \tag{6.4}$$

when $R_4 \ll R_5$. The value of R_4 should be chosen to provide a maximum set current of approximately $100 \, \mu A$ during the sampling period. This maximum set current will optimize the high-frequency performance characteristics of the buffer amplifier and achieve the maximum slew-rate ($6 \, V/\mu s$) and bandwidth ($10 \, MHz$) conditions.

The hold period is initiated when the control voltage is negative. During this period, the JFET switch Q_2 is turned off, blocking the input signal from reaching the buffer amplifier. Resistor R_5 keeps the programmable amplifier "on" but at a much reduced set-current level, to optimize the amplifier for good dc performance. That is, during the hold period, the major decay of the signal level is due to the leakage current associated with the input-bias current of the amplifier. Thus reducing the set-current level will minimize bias current, resulting in negligible voltage decay. In the sample-and-hold application, the programmable amplifier offers the capability of combining excellent ac and dc performance.

Another useful application of the current-controlled programmable amplifier is as a voltage-controlled oscillator (VCO). The VCO is a particularly valuable functional block in closed-loop telemetry systems where landline analog data are converted to an FM signal for transmission over a coaxial line. Phase-locked loops, which are investigated in Chapter 9, incorporate VCOs, as do many frequency generators.

The programmable characteristics of the HA-2720 can be used to form a controlled oscillator having a linear operation over the wide dynamic range of 1000:1 without requiring external timing capacitors. Unlike most other circuits that require external integrating capacitors, the simple circuit using two programmable amplifiers (Figure 6.6) relies on the internal compensation capacitor (C_c) for integration. Varying the set current will proportionally change the magnitude of current that is available to charge C_c, effectively altering the frequency of operation. Hence this circuit is better identified as a current-controlled oscillator (CCO).

The CCO circuit is made up of two programmable op amps on the same monolithic chip. Op amp 1 is used as a voltage follower; the second op amp functions as a comparator. The voltage follower is confined to operate in a slewing mode, which means that the maximum rate at which the output will follow the input is determined by the slew rate of op amp 1. The comparator is preset by R_4 for the maximum set current, and maximum slew rate results, thereby assuring that the voltage follower slews throughout the range. The positive-feedback circuit, consisting of R_1, R_2, and R_3, sets up the hysteresis characteristic, and its dead band is controllable by V_{ref}. The output assumes either of two states +15 V, or −15 V, and is triggered when the input ramp generated by the voltage follower exceeds V_{ref}. If the comparator output is at +15 V, the reference voltage (V_{ref}) becomes +5 V and +10 V is applied to the voltage follower. The follower output voltage (v_T) slews at the programmed slew rate established by (I_{set}), until v_T becomes slightly larger than +5 V. At this time the comparator will switch its output to −15 V, applying a reference voltage of −5 V to the comparator and −10 V at the voltage follower input. The output v_T will now slew in the negative direction until the −5 V reference voltage is exceeded. The process repeats itself at a frequency established by

$$f = \frac{SR}{4|V_{ref}|} = \frac{K(I_{set})_1}{4|V_{ref}|} = K_1(I_{set})_1 \tag{6.5}$$

This linear relationship exists from about 50 Hz to 50 kHz and is depicted in Figure 6.6b. For linear operation, the comparator slew rate must be much greater than that of the voltage follower (assuring a slew limit

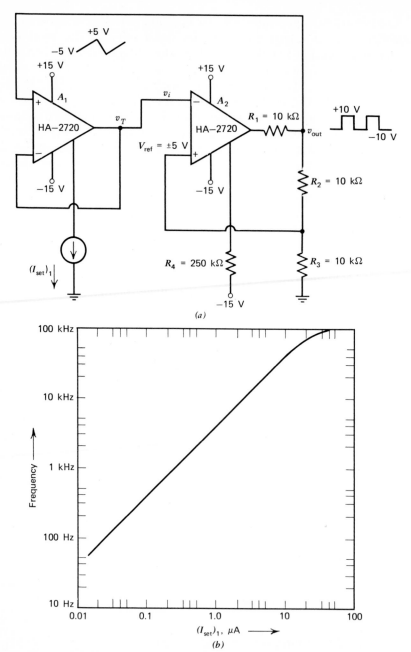

Figure 6.6 Controlled oscillator whose frequency is determined by the set current of a programmable amplifier. (a) Circuit diagram. (b) Output frequency versus set current.

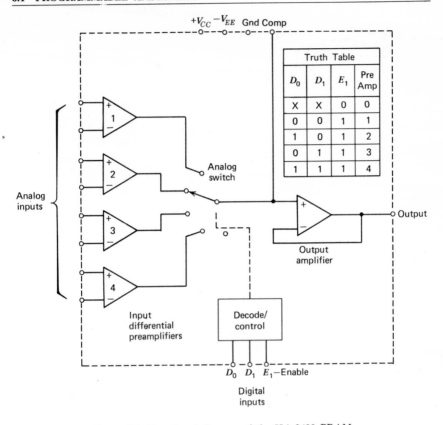

Figure 6.7 Functional diagram of the HA-2400 PRAM.

condition), so that the linearity is solely dependent on the slew-rate–I_{set} relationship. The nonlinear relationship beyond 50 kHz is due to the voltage follower's slew rate approaching that of the comparator.

A second type of programmable amplifier combines the functions of a high-performance op amp with an analog switch to narrow a bit more the gap between the digital and analog worlds. The amplifier, known as a PRAM,* is the HA-2400. The device (see Figure 6.7) contains four preamplifier sections, shown as channels 1 through 4, which are singly selectable through an analog switch that is controlled by a 2-bit binary DTL- and TTL-compatible decode input. The selected preamplifier drives an output amplifier to form an overall, high-performance op amp. In actuality, the complete circuit (Figure 6.8) consists of four conventional op amp input curcuits connected in parallel to a conventional op amp

_____

*PRAM is a trademark of Harris semiconductor for programmable amplifier.

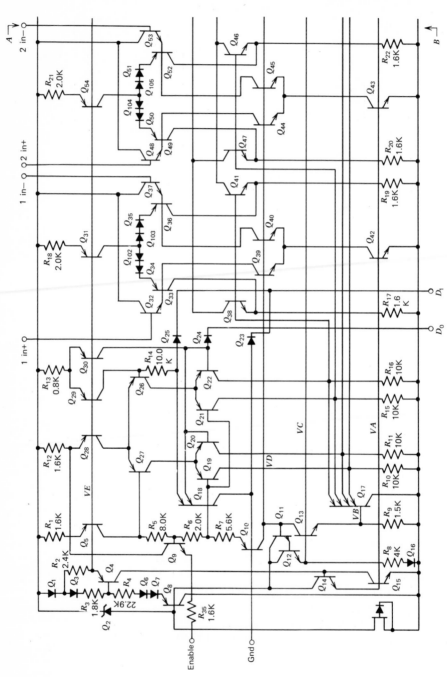

Figure 6.8 Circuit diagram for a programmable amplifier (PRAM HA-2400).

230

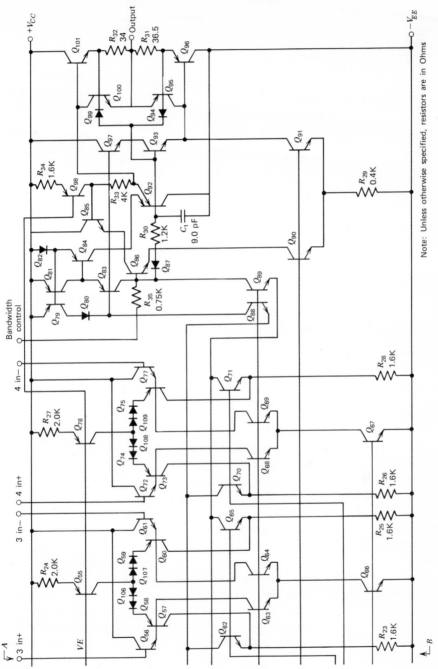

Figure 6.8 (cont.)

231

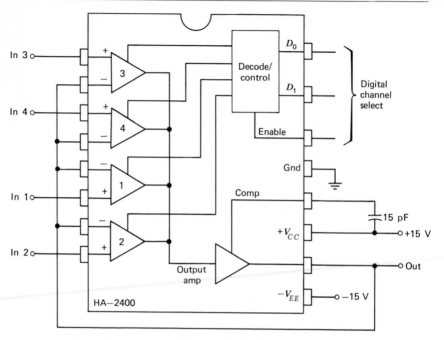

Figure 6.9 Analog multiplexer with buffered input and output.

output circuit. The decode/control circuitry furnishes operating current only to the selected input section.

The four-channel PRAM makes practical a large number of linear circuit applications. Any circuit function possible with a conventional op amp can be implemented using any one channel of the PRAM. Similar or different feedback networks can be employed around the individual channels to select and condition different input signals or to select between different op amp functions to be performed on a single input. To design an op amp function, the appropriate network is connected between the common output and the two inputs for the particular channel. It is often possible to design circuits with fewer external components than would be required for implementing four separate op amps. We now examine several applications well suited for implementation using the PRAM.

A four-channel analog multiplexer can be generated using the circuit drawn in Figure 6.9. The design uses unity feedback between the output and all four preamplifier inputs to achieve the voltage-follower configuration. Hence each channel has a very high input impedance (typically >30 MΩ) and a very low output impedance (typically <1 Ω). The single

PRAM replaces four input buffer amplifiers, four analog switches with the associated decoding, and one output buffer amplifier.

Expansion to multiplex 5 to 12 channels can be accomplished using two or three PRAMS. The compensation pins should be connected together, and the output of only one of the devices is used. The enable input must be kept low on the unselected amplifiers. Expansion to 16 or more channels is accomplished in a straightforward manner by connecting outputs of four 4-channel multiplexers to the inputs of another 4-channel multiplexer. Differential input signals can be handled by using two identical PRAMs addressed in parallel.

For low-level input signals, gain can be added to one or more channels by connecting the negative inputs to a voltage divider between output and ground. However, the bandwidth will be reduced from the voltage-follower configuration by approximately the same factor as the gain employed.

A noninverting amplifier with programmable gain appears in Figure 6.10. The feedback network employs resistors chosen to produce a gain of 0, 1, 2, 4, or 8, depending on the preamplifier channel selected. Comparators at the output could be used for autoranging meters and similar applications.

An interesting extension of the programmable gain amplifier is achieved by cascading two PRAMs together in such a way that 16 analog gains occur in unit steps (i.e., 1, 2, 3, . . . , 15, 16). The preamplifier channels utilized are so chosen that a 1:1 correspondence exists between the digital code and the total gain. For example, an input binary code of 0000 would yield a gain of +1; a code of 0001 would produce a gain of +2; and so on, up to a code of 1111, which yields a gain of +16.

The block diagram of Figure 6.11a is a scheme that can produce the 16 unit-step gains. The amplifier A represents any one preamplifier from the first PRAM, and amplifier B is any one preamplifier from the second PRAM. After signal summation, the overall gain can be expressed as

$$\frac{v_{\text{out}}}{v_{\text{in}}} = -(A + B) \tag{6.6}$$

By choosing $A = \begin{Bmatrix} -1 \\ -5 \\ -9 \\ -13 \end{Bmatrix}$ and $B = \begin{Bmatrix} 0 \\ -1 \\ -2 \\ -3 \end{Bmatrix}$, all gains of 1 through 16 are possible. The circuit of Figure 6.11b illustrates the general interconnection scheme employed between one op amp from the first PRAM and one op amp from the second PRAM. The desired overall gain of $-(A + B)$ is

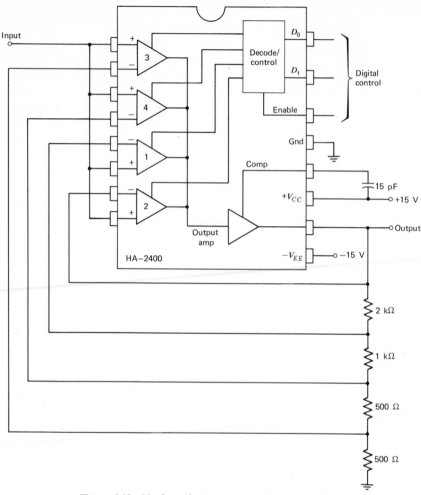

Figure 6.10 Noninverting programmable gain amplifier.

set through the six feedback resistors according to

$$A = -\frac{R_2 R_4}{R_1 R_3} \tag{6.7}$$

and

$$B = -\frac{R_6(R_3 + R_4)}{R_3(R_5 + R_6)} \tag{6.8}$$

Figure 6.12 shows the complete circuit interconnections between the two PRAMs which achieve the desired 16-step gains. Different amplifier combinations for the signal path are selected via the decode control

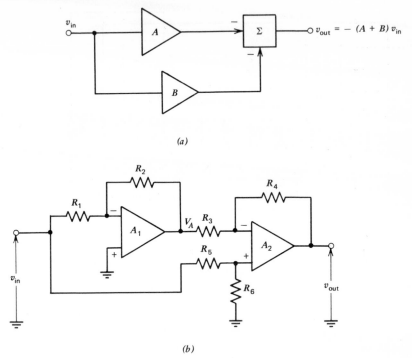

(a)

(b)

Figure 6.11 Programmable gain amplifier. (a) Block diagram of design approach. (b) Op amp interconnection scheme.

inputs. (Figure 6.13 gives the individual amplifiers selected and their associated gains which achieve the desired total gain.) The four resisitors connected to the noninverting terminals of the first PRAM and the two 667 Ω resistors connected to the second PRAM serve only for input-bias current compensation and do not affect the gain calculations.

The scheme depicted considerably enhances the versatility of a single 4-channel PRAM and introduces a number of interesting applications such as digital automatic gain control (AGC), easier digital control of servo systems, and level detectors, just to cite a few.

The PRAM can be programmed to become an analog signal adder-subtractor using the circuit of Figure 6.14. The output can be a function of the two inputs, x and y, in any one of the following ways:

$$v_{out} = \begin{cases} -K_1 x & \longrightarrow \text{ channel 1} \\ -K_2 y & \longrightarrow \text{ channel 2} \\ K_3 x - K_4 y & \longrightarrow \text{ channel 3} \\ -(K_5 x + K_6 y) & \longrightarrow \text{ channel 4} \end{cases} \qquad (6.9)$$

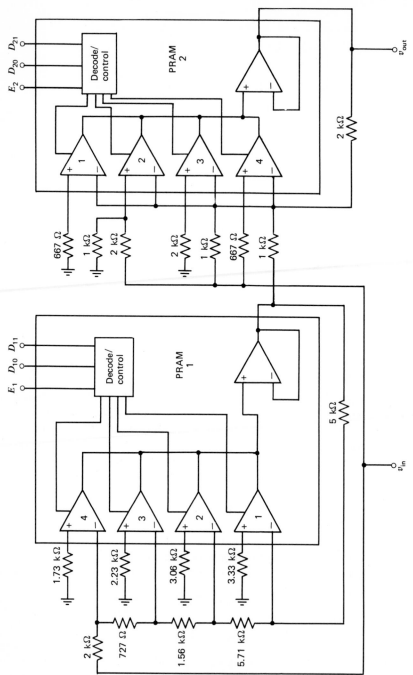

Figure 6.12 Circuit diagram for 16-step programmable amplifier.

Control				Enable		PRAM 1		PRAM 2		Total Gain
D_{10}	D_{11}	D_{20}	D_{21}	E_1	E_2	Amplifier	Gain A	Amplifier	Gain B	
$\times^a$	$\times$	$\times$	$\times$	$\times$	0^b	None	—	None	0	0
0	0	0	0	1	1	1	−1	1	0	1
0	0	0	1	1	1	1	−1	2	−1	2
0	0	1	0	1	1	1	−1	3	−2	3
0	0	1	1	1	1	1	−1	4	−3	4
0	1	0	0	1	1	2	−5	1	0	5
0	1	0	1	1	1	2	−5	2	−1	6
0	1	1	0	1	1	2	−5	3	−2	7
0	1	1	1	1	1	2	−5	4	−3	8
1	0	0	0	1	1	3	−9	1	0	9
1	0	0	1	1	1	3	−9	2	−1	10
1	0	1	0	1	1	3	−9	3	−2	11
1	0	1	1	1	1	3	−9	4	−3	12
1	1	0	0	1	1	4	−13	1	0	13
1	1	0	1	1	1	4	−13	2	−1	14
1	1	1	0	1	1	4	−13	3	−2	15
1	1	1	1	1	1	4	−13	4	−3	16

a Don't care.
b Disenabling PRAM 2 prevents signal summation, resulting in zero total gain.

Figure 6.13 Truth table for the 16-step programmable gain amplifier.

where the K's are all positive constants determined by the feedback resistors. Obviously, many other functions of one or more variables could be generated from the basic design approach illustrated.

Any oscillator that can be constructed using an op amp (e.g., the twin-T, phase-shift, and crystal-controlled types) can be made programmable by using the PRAM. Illustrated in Figure 6.15 is a Wien bridge type, which is very popular for signal generators, since it is easily tunable over a wide frequency range and has a very-low-distortion sine-wave output. The RC frequency-determining networks can be designed from about 10 Hz to greater than 1 MHz. The output level is about 6.0 V rms. By substituting a programmable attenuator for the buffer amplifier, a very versatile sine-wave source for automatic testing, and other purposes can be constructed.

Application Hints

It should be noted when using the PRAM that the feedback networks for the unused channels can still constitute a load at the amplifier output and

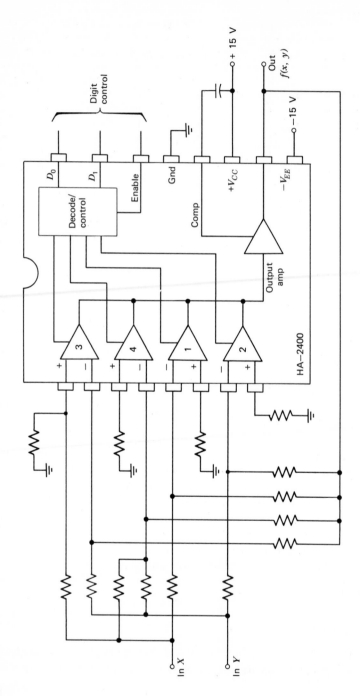

Figure 6.14 Adder-subtractor programmable function.

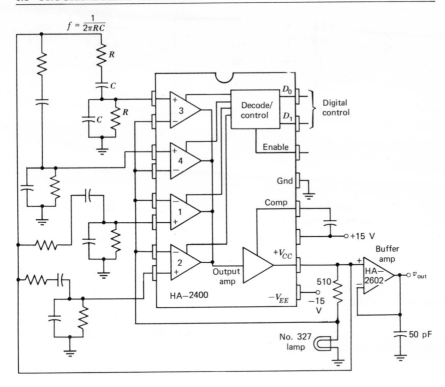

Figure 6.15 Sine-wave oscillator having a programmable frequency capability.

at the signal input as if the unselected input terminals were disconnected. This is particularly significant in inverting-amplifier connections, where the feedback resistors may cause crosstalk from the output to unselected inputs.

Another important question is, What happens to the PRAM output when the enable pin is a logic 0? In this situation, all four preamplifiers are disabled. The output amplifier, which functions as a voltage follower, now has a floating input. Normally in this mode, the bias current of the output amplifier slowly discharges the compensation capacitor and the output eventually goes toward the negative power supply. This situation is seldom critical when the coupling is ac, since all input signals are blocked from the output. However, the large negative output can seriously degrade dc applications.

The easiest solution is to utilize an unused channel as a voltage follower with its input grounded. When this channel is addressed, the output voltage will be within a few millivolts of zero volts. If all channels are being used, some extra components must be added to force the output to

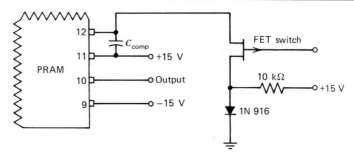

Figure 6.16 Auxiliary circuit for forcing PRAM output to zero volts when enable input is low.

zero. Such a connection is shown in Figure 6.16. The FET switch connected to the compensation pin forces the output to zero when the FET is turned on with a logic 0 input. Since the output voltage will be one forward diode drop (about 0.7 V) more negative than the voltage at the compensation pin, a diode in series with the switch is necessary. The FET switch should have very low "off" leakage current, but it will normally carry less than 1 μA when "on," provided the enable input is also a logic 0. (Adding a second PRAM with its compensation pin connected to that of the first and using the voltage-follower connection is probably an easier and cheaper method of obtaining zero output.)

6.2 CHOPPER-STABILIZED AMPLIFIERS

For about 30 years, the chopper-stabilized amplifier (chopper) has been employed only when an advanced amplifier with the ultimate in dc performance was required. Initially the chopper was constructed with vacuum tubes and mechanical relay switches or choppers. More recently, this amplifier has been made with discrete solid-state devices in a module or hybrid package. Now a monolithic chopper has been fabricated which achieves the same performance with compactness, higher reliability, and lower cost (see Figure 1.2).

Probably the one most troublesome parameter to the monolithic operational amplifier designer is offset-voltage drift. This is also the most basic parameter for a dc amplifier—with zero volts differential input, the amplifier ought to have zero volts output. Input-offset voltage (the small voltage that would be required between the *input* terminals to make the output actually go to zero volts) can generally be adjusted to a very low value using an external potentiometer or selected fixed resistors. Unfortunately, this adjustment is good only for the ambient temperature (and instant of time) at which the adjustment is made. With a few degrees of change in temperature or after a few months, the offset voltage may again become significant. The chopper-stabilized amplifier is designed to reduce the offset-voltage and current-drift to levels so small that they are virtually unmeasurable by conventional methods. We first discuss the methods incorporated in the design of a chopper to achieve the improved dc performance. Following this discussion, we show several applications well suited for the utilization of the chopper-stabilized amplifier.

Stabilization of the offset voltage can be accomplished by using a two-amplifier system like that in Figure 6.17a. In this figure, A_1 is the main amplifier and A_2 is an auxiliary high-performance dc amplifier, which has a very limited frequency response but a very low offset-voltage drift. The voltage gain of the two-amplifier system is

$$A_v = A_1(1 + A_2) \approx A_1 A_2 \qquad (6.10)$$

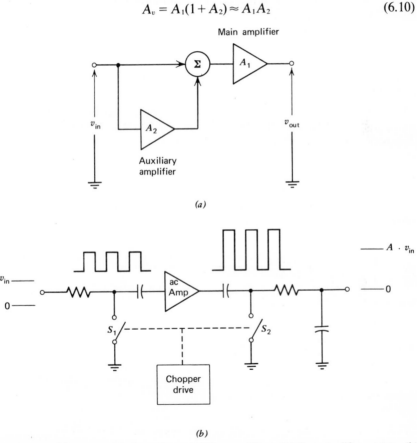

(a)

(b)

Figure 6.17 Common techniques for stabilizing amplifiers. (a) Two-amplifier system. (b) Chopper stabilization. (Reprinted courtesy of *Electronic Magazine*.)

If the gain of A_2 is large, the effective input-offset voltage of the entire circuit will be nearly that of A_2 alone. This is because the input-offset voltage of A_1 is effectively divided by the gain of A_2 in determining its contribution to the offset of the entire circuit. As Equation 6.10 shows, the dc gain of the entire circuit is the product of the gains of A_1 and A_2.

The classical chopper amplifier (Figure 6.17b) is often used as the auxiliary amplifier (A_2) in a stabilized configuration. The chopper switch S_1 functions as a modulator that changes the incoming dc level to an ac wave form having a proportional amplitude, and phase angle of either 0 or 180°, depending on input polarity. The chopped signal is then amplified by an ac-coupled amplifier. Ground level of the amplified signal is restored by a second chopper switch S_2, which may be regarded as a synchronous demodulator. Filtering then recreates an amplified replica of the incoming dc or low-frequency signal.

This circuit, properly constructed, will have extremely low offset-voltage drift. The amplifier, being ac coupled, does not contribute to the dc offset. The most critical element is the switch S_1, since any coupling, dc or ac, of the drive signal to the contacts can introduce an offset error.

A different chopper-amplifier concept (Figure 6.18) is a dc-coupled scheme in which the amplifier periodically disconnects itself from the input signal and adjusts its offset voltage to zero. With all switches in the X position, the circuit functions as a dc amplifier. When all switches go to the Y position, the amplifier input is grounded, and A_2 forces the output of A_1 to ground. Switch S_2 and C_1 form a sample-and-hold configuration, permitting the correction signal to be stored on C_1 after S_2 returns to position X to zero the offset of A_1. Switch S_3, C_2, and A_3 form a second sample-and-hold configuration, whose function is to store the previous

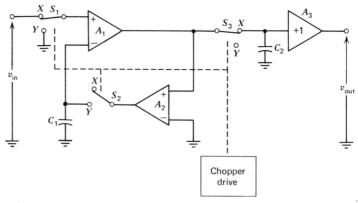

Figure 6.18 Self-zeroing amplifier. (Reprinted courtesy of *Electronics Magazine*.)

output of A_1. This occurs while self-zeroing is taking place, thereby removing most of the signal discontinuity.

When monolithically integrating the chopper amplifier, the second scheme in Figure 6.18 is more attractive, although the block diagram seems to be more complex. This scheme uses small-value resistors and fewer external capacitors, all having one end grounded. Furthermore, the scheme has performance advantages. The absence of coupling capacitors provides much faster recovery from saturation conditions—a notorious problem with traditional chopper-stabilized amplifiers. The response of the sample-and-hold filter is flat to one-half the chopper frequency, which greatly reduces settling times. Finally, this scheme may be readily modified to provide a stabilized amplifier with full differential inputs—a highly desirable feature.

In the block diagram of the monolithic chopper-stabilized amplifier (Figure 6.19), the A_1 block is the main amplifier, and A_2 is the auxiliary stabilizing amplifier. The third A_3 is the sample-and-hold amplifier in the self-zeroing loop of A_2, and A_4 is the sample-and-hold amplifier that holds the previous signal during the zeroing interval.

The design approach of Figure 6.19 and those previously discussed clearly differ in that the input circuitry in Figure 6.19 is completely symmetrical with respect to the two input lines. This produces a true differential input, in contrast to most stabilized amplifiers, which are designed either as inverting-only or as noninverting amplifiers.

During the period in which A_2 is stabilizing A_1, switches S_1 and S_4 are closed while S_2 and S_3 are open, as the timing diagram shows. The dc and low-frequency components of the input are amplified by A_2 and applied as a correction signal to A_1. The effective input-offset voltage is nearly that of A_2 alone.

The offset voltage of A_2 is kept extremely low by periodic zeroing. When S_1 opens, S_2 closes and disconnects A_2 from the input terminals and shorts the inputs of A_2 together—not at ground level, but to a level equal to the input common-mode voltage. This results in an extremely high common-mode rejection ratio.

Like most other monolithic op amps, this device does not have a ground terminal; thus when S_3 closes, the output of A_2 is forced to be equal to an internally generated reference voltage, rather than to ground. Since A_4 is referenced to the same voltage, the result is the same. Capacitor C_2 charges to a level that will maintain the offset voltage of A_2 at zero. In the meantime, S_4 has opened, which means that C_1 maintains its previous level. The offset of A_2 has now been zeroed, and A_2 returns to its task of stabilizing A_1.

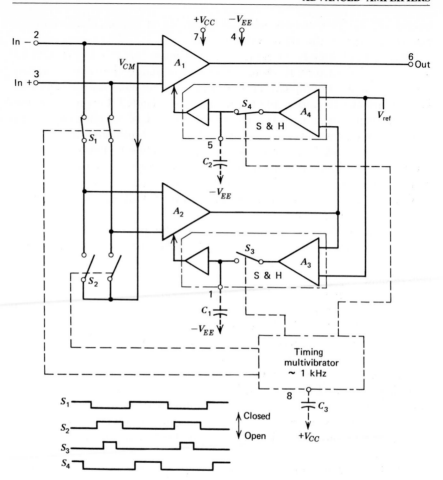

Figure 6.19 Block diagram of monolithic chopper-stabilized amplifier (HA-2900). (Reprinted courtesy of *Electronics Magazine*.)

Note that the opening and closing times of S_1 through S_4 are interleaved. This allows the transient spikes generated when a switch is opened or closed to settle out before the actuation of other signal paths, which could be affected by these transients.

A timing multivibrator (Figure 6.20) generates a triangular wave form. Different levels of the triangular wave form are detected by the four comparator circuits, which are referenced to different points of a voltage divider to produce the four desired, switch-driving signals. Analog switches S_1 and S_2 are each composed of two N-channel MOS FETs, which make excellent choppers with no associated offset voltages for the

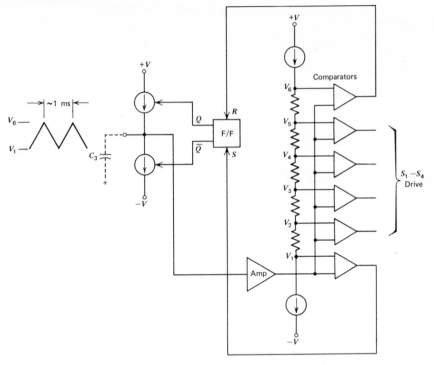

Figure 6.20   Multivibrator and switch drivers.

low levels and low currents involved. Switches S_3 and S_4 are complementary bipolar current switches, since appreciable current drive is required and offset voltage is not critical at these points. The switch circuits appear in Figure 6.21. Amplifiers A_1 and A_2 are each N-channel MOS FET input amplifiers, which produce the extremely low input currents (Figure 6.22). Normally, MOS FETs would not be suitable as dc amplifier-input stages because of their high offset-voltage drift; but chopper stabilization effectively removes that drift, while retaining the high-input impedance advantage associated with MOS FETs.

Single-ended MOS FET input stages are used in the two sample-and-hold circuits as buffers to sense capacitor voltages. The correction signal from each sample-and-hold circuit alters a current generator that feeds one of the MOS FET sources in the inputs of A_1 and A_2. The output stage of A_1 is a conventional complementary bipolar follower with short-circuit protection.

All this complex circuitry is incorporated into a simple functional op amp block, shown in Figure 6.23 packaged with the standard op amp pinout in the standard TO-99 can. Three external capacitors are required

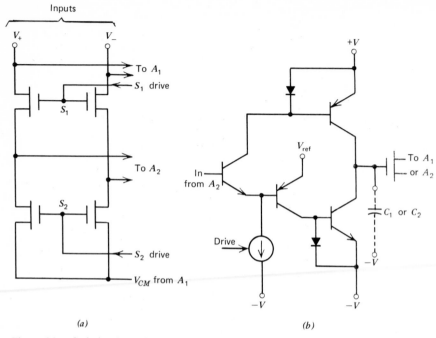

(a)

(b)

Figure 6.21 Switch schematics. (a) Switches S_1 and S_2. (b) Sample-and-hold amplifier and switch.

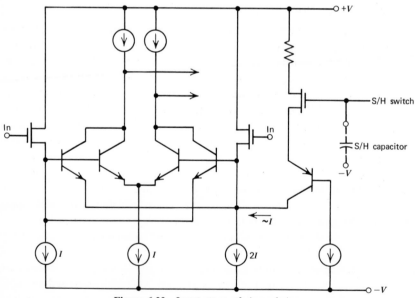

Figure 6.22 Input stage of A_1 and A_2.

246

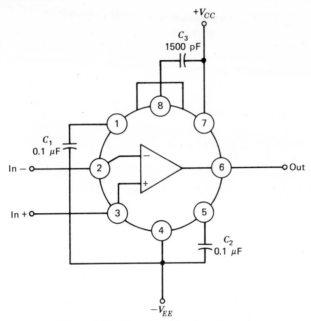

Figure 6.23 Pinout and hookup diagram.

for operation: one for multivibrator timing and two for the sample-and-hold.

The fabrication of the monolithic chopper-stabilized amplifier represents a large-scale integration linear device. The chip contains 252 active elements on a die measuring 0.093×0.123 in.

The chip was designed using the dielectric isolation process, rather than the more conventional junction isolation process, for several reasons. An analog chip can usually be made smaller in dielectric isolation, both because of better packing density and because the high-quality active elements can simplify the design, requiring fewer high-value resistors and capacitors. The savings in chip area and the consequent higher yield mean that dielectric isolation can be used at little or no cost premium.

The factor that makes a monolithic chopper-stabilized amplifier genuinely practical is the high-quality NPN, PNP, and FET elements that can be readily fabricated using dielectric isolation. The circuit designer has more freedom in the choice of transistors with desirable parameters, such as high-frequency, high-gain, vertical PNP transistors, and MOS FETs optimized for chopper service. The superior isolation between elements greatly reduces parasitic capacitance and prevents interaction or latchup due to unwanted four-layer devices. This also allows accurate circuit modeling, which is essential in a circuit of this complexity.

Typical Parameters over -55 to $+125°C$

Parameter	HA-2900	Units
Offset voltage at $+25°C$	20	μV
Offset voltage drift	0.3	$\mu V/°C$
Offset current at $+25°C$	0.05	nA
Offset current drift	1.0	pA/°C
Open-loop gain	5×10^8	V/V
Bandwidth (for unity gain)	3	MHz
Slew rate (unity gain)	2.5	$V/\mu s$

Figure 6.24 Parameters for low-drift monolithic chopper-stabilized op amp.

Furthermore, thermal symmetry is an important consideration in circuit layout, to prevent the heat generated by either output transistor from producing thermal gradients between comparable input transistors.

Significant performance parameters of the chopper are listed in Figure 6.24. (These parameters should be compared with the typical op amp parameters previously cited in Figure 5.1.) In addition to the superior dc parameters, the bandwidth and slew rate are quite respectable. The amplifier is stable, even as a unity-gain follower, and exhibits a smooth, fast-settling slewing wave form (see Figure 6.25).

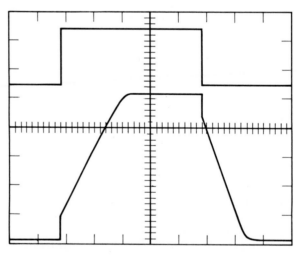

Upper trace: input; 5 V/division
Lower trace: output; 2V/division
Horizontal: 2 μs/division
$R_L = 2$ kΩ; $C_L = 50$ pF

Figure 6.25 Voltage-follower slewing wave form.

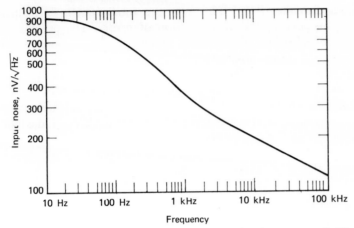

Figure 6.26 Input voltage noise characteristics. (Reprinted courtesy of *Electronics Magazine.*)

One other factor to be considered is the amplifier's equivalent input noise. As is generally true of most other chopper-stabilized amplifiers, noise voltage is several times higher than most nonstabilized amplifiers. In the HA-2900, however, this noise is not due to the chopper action. Figure 6.26 indicates that a $1/f$ type of random-voltage noise is predominant, and the noise is higher than usual because of the extremely high impedance of the MOS FET input stage.

Synchronous noise generated by the choppers is primarily a common-mode current noise, which can be minimized by matching the impedances at the two input terminals. With matched impedances up to 100 kΩ, the synchronous noise seen at the output is well below the random noise level; and the effect of random-current noise is not discernible at this impedance level.

Applications particularly well suited for choppers include precise analog computation and precision dc instrumentation where low drift is an outstanding requirement.

Applications in which the need for a chopper is less obvious can be determined by careful consideration of the following points.

Assemblies that require periodic adjustment of a potentiometer or selected resistors to zero an amplifier are well suited for utilization of a chopper. Cost and reliability advantages usually favor using an amplifier that requires no circuit adjustments. This is especially true of systems calling for occasional recalibration because of amplifier drift with time. Designs that must be reworked because of excessive amplifier drift with temperature are particularly well suited for a chopper because the drift specification in many amplifiers may no longer be valid after zeroing.

Choppers also are useful whenever the total system performance is marginal because of the accumulation of errors. The designer should consider whether the error-budget situation would improve with the substitution of much more accurate op amps. A complex analog/digital system may be avoided by using the chopper to improve the accuracy and drift problems associated with a simpler all-analog system.

Whether a chopper-stabilized amplifier is used in such cases will depend on analysis of the cost and performance tradeoffs in the individual situations. In any case, the knowledge that a solution is now available, should any of these problems arise, will remove some of the greatest worries of the linear systems designer.

Applications of Chopper-Stabilized Amplifiers

Some care is required in the physical layout of the system to realize the full accuracy potential of an ultra-low-drift amplifier. When mounted in a typical breadboard or printed-circuit card adequate for an ordinary op amp application, drifts on the order of $1 \mu V/°C$ may be expected. If this is good enough, the designer need go no further. But to reach the ultimate device performance, the designer must take into account external effects. These include thermocouple and electrochemical EMFs generated at junctions of dissimilar metals (solder points, connectors, internal junctions in resistors and capacitors), leakage across insulating materials,

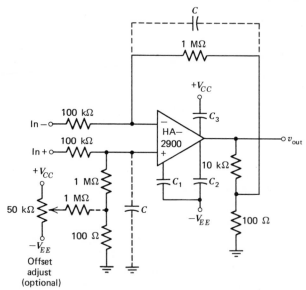

Figure 6.27   Amplifier with gain of 1000. (Reprinted courtesy of *Electronics Magazine*.)

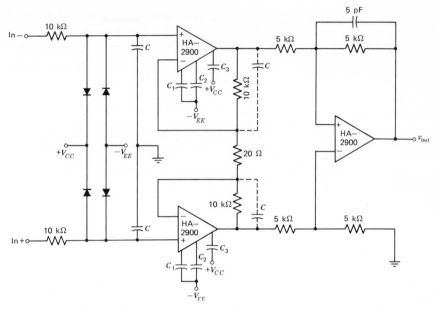

Figure 6.28 Differential instrumentation amplifier. (Reprinted courtesy of *Electronics Magazine.*)

static charges created by moving air, and improper grounding and shielding practices. The main layout procedure is to ensure that the networks going to the two amplifier inputs are identical and are at the same temperature.

In the typical high-gain amplifier application of Figure 6.27, the gain is 1000 and the bandwidth is about 2 kHz. Either input terminal may be grounded for inverting or noninverting operation, or the inputs may be driven differentially. The symmetrical networks at the device inputs are recommended for any of the three operating modes to eliminate chopper noise and to yield the best drift characteristics. Total input noise, with $C = 0$, is about 30 μV rms. This noise can be reduced, at the expense of bandwidth, by adding capacitors as shown.

A high-input impedance differential instrumentation amplifier is presented in Figure 6.28. This well-known configuration has common-mode rejection for over ± 10 V common-mode input signals. Protection diodes are included to prevent the device input voltages from exceeding either power supply.

Analog integrators have long been utilized in analog computation, active filters, timers, wave-form generators, control systems, and A/D converters. An op amp for a precision integrator should have high gain, low offset voltage, low bias current, and wide bandwidth. Therefore, the

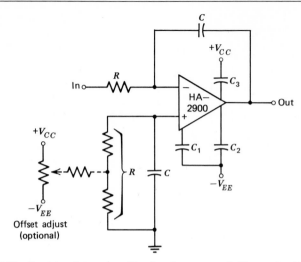

Figure 6.29 Precision integrator. (Reprinted courtesy of *Electronics Magazine.*)

chopper-stabilized amplifier is well suited as the active device in an integrator like that in Figure 6.29. The high gain of the chopper allows accurate integration over eight decades of frequency. Dual-slope A/D converters can now easily be made with six-digit resolution.

Figure 6.30 illustrates one of several methods of producing a composite amplifier that combines the outstanding dc characteristics of the chopper with the ac characteristics of a wide-band and/or high-slew-rate amplifier.

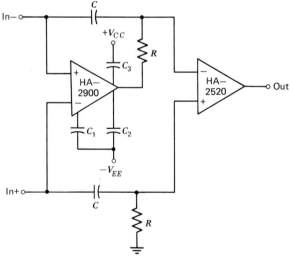

Figure 6.30 Composite high-slew-rate, low-drift, amplifier. (Reprinted courtesy of *Electronics Magazine.*)

One application for this type of circuit is as a buffer amplifier for a fast, high-accuracy D/A converter. The combined frequency response of the circuit may be tailored for optimum settling time within a very narrow error band.

6.3 SAMPLE-AND-HOLD GATED OPERATIONAL AMPLIFIERS

The sample-and-hold (S/H) function is a very widely used operation in linear systems. An S/H device has an analog signal input, a control input, and an analog signal output. The device always operates in one of its two modes: SAMPLE, in which it acquires the input signal as rapidly as possible and tracks it faithfully until commanded to HOLD, at which time it retains the last value of the input signal that it had at the time the control signal called for a mode change. Stated more simply, the S/H circuit is an analog switch which, on command, samples the instantaneous level of an analog signal and retains this voltage as a dc level. Figure 6.31 shows typical operational wave shapes for the S/H function.

Sample-and-hold devices can be thought of as a pair of gated op amps, normally connected in the voltage-follower configuration. The control input is normally operated by "standard" logic levels and is usually TTL or MOS compatible. Until recently, the S/H function was available only in modular or hybrid circuits. More frequently, however, the circuit has been constructed by the user from an analog switch, a capacitor, and a very-low-bias-current op amp, as in Figure 6.32.

A high-quality sample-and-hold circuit must meet three specific requirements:

1. The holding capacitor must charge up and settle to its final value as quickly as possible.
2. When holding, the leakage current at the capacitor must be as near zero as possible, to minimize voltage drift with time.
3. Other sources of error must be minimized.

Design of a sample-and-hold circuit, particularly the user-built variety, involves a number of compromises in the requirements just listed. The amplifier or other device feeding the analog switch must have high-current capability and must be able to drive capacitive loads with stability. The analog switch must have both low "on" resistance and extremely low "off" leakage currents. However, the leakage currents of most analog switches (except the dielectrically isolated types) tend to run several hundred nanoamperes at elevated temperatures. The analog

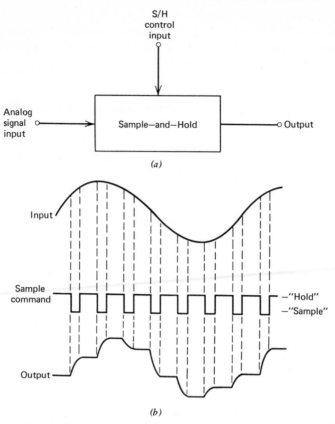

Figure 6.31   The sample-and-hold device. (*a*) Block-diagram representation. (*b*) Typical S/H wave forms.

switch must have very low coupling between the digital input and analog output, because any spikes generated at the instant of turn-off will change the charge on the capacitor. The output amplifier must have extremely low bias current over the temperature range, as well as low offset drift and sufficient slew rate—a combination satisfied by only a few available amplifiers.

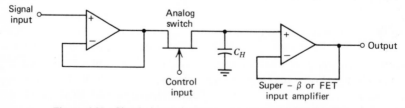

Figure 6.32   Simple implementation of sample-and-hold device.

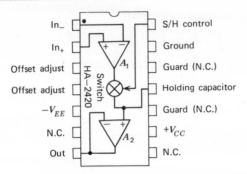

Figure 6.33 Functional diagram of monolithic sample-and-hold device.

In the first monolithic AIC S/H device (Figure 6.33), the input amplifier stage (A_1) is a high-performance operational amplifier having excellent slew rate and the ability to drive high-capacitance loads without instability. The switching element is a highly efficient bipolar transistor stage with extremely low leakage in the "off" condition. The output amplifier (A_2) is a MOS FET input unity-gain follower, chosen to achieve extremely low bias current.

MOS FET inputs are seldom used for dc amplifiers because their offset voltage drift is difficult to control. In this configuration, however, negative feedback is generally externally applied between the output and inputs of the entire device, and the effect of this offset drift at the inputs is divided by the open-loop gain of the input amplifier stage.

The schematic of the monolithic S/H device appears in Figure 6.34. During sampling (S/H control "low") the signal path through the input amplifier stage starts at Q_{31}–Q_{34}, through Q_{45} and Q_{46}, and then to the holding capacitor terminal through Q_{51}–Q_{54}. The output-follower amplifier has its input at MOS FET Q_{60}.

In the "hold" mode, the S/H control is "high," so Q_{21} conducts, turning on Q_{27} which diverts the signal away from Q_{45} and Q_{46}, and passes the signal to $-V_{EE}$ through Q_{57}. Since Q_{57} also forces Q_{51}–Q_{54} to ride up and down with the output signal, there is virtually zero potential between these transistor bases and the voltage on C_H, completely eliminating leakage from C_H back into the input amplifier.

Sample-and-Hold Applications

We now examine several important applications of the S/H circuit to illustrate the versatility of this monolithic device. An op amp having a highly efficient analog switch in series with its output is a very useful building block for linear systems because virtually any of the hundreds of

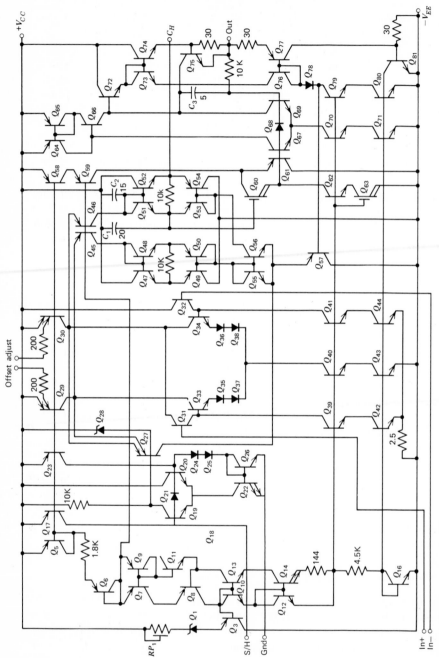

Figure 6.34 Sample-and-hold circuit diagram. Unless otherwise specified, resistance values are in ohms, capacitance values are in picofarads.

feedback configurations can be implemented. As Figure 6.33 indicates, the S/H functions as a conventioned op amp when the control switch is closed. In this mode, the op amp has excellent bandwidth, slew rate, and high-output current capability, and is able to drive capacitive loads with good stability. With the switch open, the output mode is almost a perfect open circuit. The output buffer amplifier has excellent, high input impedance characteristics and exceptionally low bias current. Since its offset voltage may be quite high, however, this stage is not particularly well suited for dc applications *outside* an overall feedback loop. Let us now study several specific applications of the S/H in which this single device can be used to replace both an operational amplifier and a separate sample-and-hold module.

Implementation of the basic sample-and-hold operation is depicted in Figure 6.35. Note that the feedback connection from the output to the inverting input of the first op amp produces the conventional voltage follower, giving unity gain with a noninverted output signal. This arrangement also produces a very large input impedance.

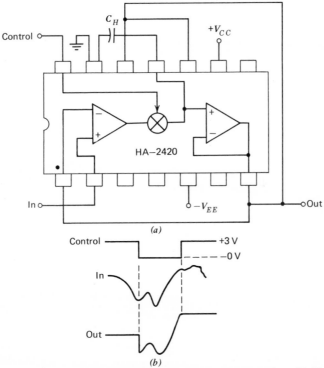

Figure 6.35 The basic sample-and-hold device. (*a*) Circuit connections (*b*) Typical wave forms.

Frequently the sample-and-hold is referred to as a track-and-hold function. The only difference between the terms is the time period during which the switch is closed. In track-and-hold operation, the switch is closed for a relatively long period during which the output signal may change appreciably; and the output will hold the level present at the instant the switch is opened. In sample-and-hold operation, the switch is closed only for the period of time necessary to fully charge the holding capacitor.

The larger the value of the timing capacitor, the longer it will hold the signal without excessive drift; however, it will also reduce the charging rate/slew rate and the amplifier bandwidth during sampling. Thus the capacitance value must be optimized for each application. Figure 6.36

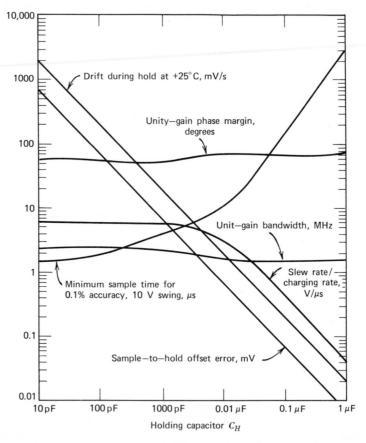

Figure 6.36 Typical sample-and-hold performance as a function of holding capacitance.

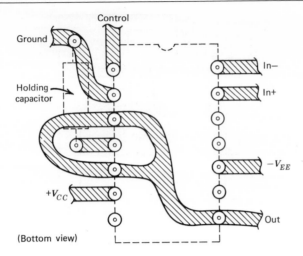

Figure 6.37 Utilization of a guard ring to reduce leakage current from the holding capacitor.

plots these tradeoffs. Drift during holding tends to double for every 10°C rise in ambient temperature. The holding capacitor should have extremely high insulation resistance and low dielectric absorption; polystyrene (below +85°C), Teflon, or mica types are recommended.

For minimum drift during holding, leakage paths on the printed-circuit board and on the device package surface must be minimized. Since the output voltage is nearly equal to the voltage on C_H, the output line may be used as a guard line surrounding the line to C_H. Then, since the potentials are nearly equal, very low leakage currents will flow. The two package pins surrounding the C_H pin are not internally connected and may be used as guard pins to reduce leakage on the package surface. A suggested printed-circuit guard-ring layout appears in Figure 6.37.

In another interesting application (Figure 6.38), the S/H device performs both op amp and sampling functions, thereby eliminating the need for a separate scaling amplifier and a sample-and-hold module. In this circuit the gain is set by the feedback resistors according to

$$G_+ = \frac{R_1 + R_2}{R_1} \tag{6.11}$$

It is usually best design practice to scale the gain such that the largest expected signal will give an output close to ±10 V. Drift current is essentially independent of output level, and smaller percentage drift will occur in a given time for a larger output signal.

An alternative S/H connection with gain is illustrated in Figure 6.39.

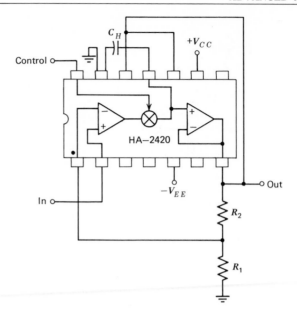

Figure 6.38 A noninverting sample-and-hold circuit with gain.

Since this is the inverting mode connection, the gain is

$$G_- = -\frac{R_2}{R_1} \qquad (6.12)$$

To minimize offsets, R_3 should be chosen according to

$$R_3 = \frac{R_1 R_2}{R_1 + R_2} \qquad (6.13)$$

The inverting connection will have somewhat greater input-to-output crosstalk in the "hold" mode, since a voltage divider is formed between $(R_1 + R_2)$ and the device output impedance. Frequently in S/H systems, signal filtering is required prior to sampling. The entire operation can be accomplished with only one device using a typical connection like that in Figure 6.40. Any of the inverting and noninverting filters that can be built with op amps can be implemented. However, the sampling switch must be closed long enough to allow the filter to settle when active filter types are connected around the device.

Short sample times require a low-value holding capacitor, whereas long, accurate hold times require a high-value holding capacitor. Thus achieving a very long hold with a short sample appears to involve contradictory effects. However, the task can be accomplished by cascading two S/H circuits, as in Figure 6.41. The first S/H circuit has a

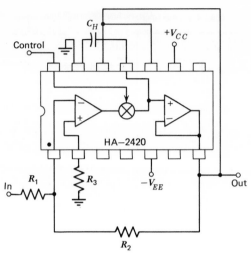

Figure 6.39 An inverting sample-and-hold circuit with gain.

low-value capacitor; the second has a high value. Then the second S/H can sample for as long as the first circuit can accurately hold the signal.

Multiplexing two or more S/H circuits is possible using the technique represented in Figure 6.42, where a common holding capacitor is shared between devices. The number of devices to be multiplexed is limited only by the requirement that the leakage currents of all devices add together, which increases the drift during holding.

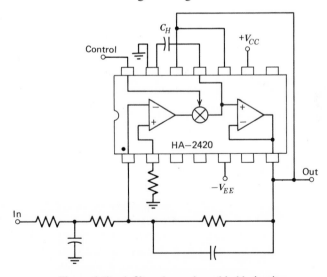

Figure 6.40 A filtered sample-and-hold circuit.

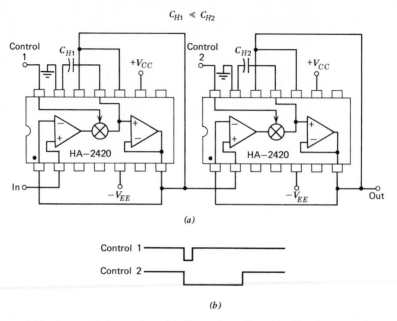

$C_{H1} \ll C_{H2}$

(a)

(b)

Figure 6.41 A cascaded sample-and-hold configuration. (a) Circuit connections. (b) Control wave shapes.

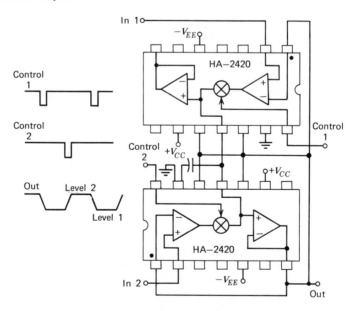

Figure 6.42 A multiplexed sample-and-hold.

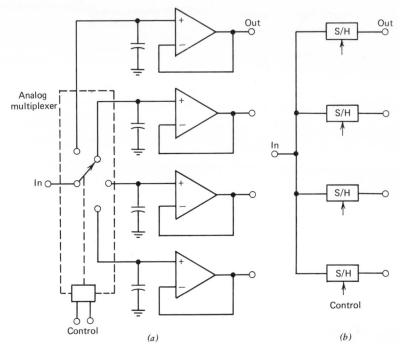

Figure 6.43 Analog demultiplexing. (*a*) Conventional technique. (*b*) Sample-and-hold approach.

The inverse multiplexing operation, or demultiplexing, reconstructs and separates analog signals that have been time-division multiplexed. Conventional demultiplexing (Figure 6.43*a*) has several restrictions, particularly when a short dwell time and a long, accurate hold time are required. The capacitors must be charged from a low-impedance source through the resistance- and current-limiting characteristics of the multiplexer. When holding, the high-impedance lines are relatively long and are subject to noise pickup and leakage. When FET input buffer amplifiers are used to ensure low leakage, severe temperature-offset errors are often introduced. The S/H configurations in Figure 6.43*b* greatly diminish all these problems.

Certain analog-to-digital converters, such as the successive-approximation type, require that the input signal be a steady dc level during the conversion cycle. The S/H device is ideal for holding the signal steady during conversion; it also functions as a buffer amplifier for the input signal, adding gain, inversion, and so on, if required.

The system illustrated in Figure 6.44 is a complete 8-bit successive-approximation converter requiring only four ICs and capable of up to

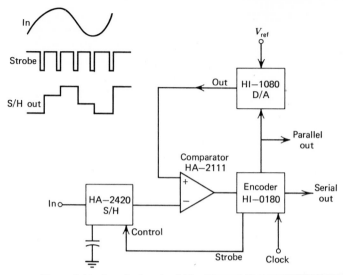

Figure 6.44 Interfacing the S/H with an A/D converter.

40,000 conversions per second. A further discussion of the A/D and D/A converters is provided in Chapter 8.

Another utilization of the S/H in analog-to-digital applications is as a "deglitcher." The word "glitch" has been universal electronics slang for an unwanted transient condition. In D/A converters, the word has achieved semiofficial status for an output transient that momentarily goes in the wrong direction when the digital input address is changed. In Figure 6.45, the S/H device does double duty, serving as a buffer amplifier as well as a glitch remover, delaying the output by one-half clock cycle. The S/H may be used to remove many other types of "glitch" in a system. If a delayed sample pulse is required, it can be generated using a dual monostable multivibrator IC.

A linear sampling system with automatic offset zeroing illustrates another application of the S/H circuit. The basic circuit in Figure 6.46 has widespread applications in instrumentation, A/D conversion, digital voltmeters (DVMs) and digital panel meters (DPMs) to eliminate offset-drift errors by periodically rezeroing the system. Basically, the input is periodically grounded and the output offset is sampled and fed back to cancel the error. The system illustrated in Figure 6.46 automatically zeros a high-gain amplifier. Care in design is necessary to assure that the zeroing loop is dynamically stable. A second sample-and-hold circuit could be added in series with the output to remove the output discontinuity. Many variations of this automatic zeroing scheme are possible to suit individual systems using the previously cited applications.

Electronic implementation of the mathematical operation of integration has always been tricky. Resetting circuits have long presented the major design problem. The reset circuit must produce an extremely low leakage current across the integrating capacitor, as well as a very low offset voltage when turned on.

The circuit appearing in Figure 6.47 accurately performs analog integration according to

$$v_{\text{out}}(t) = \int_{T_1}^{T_2} v_{\text{in}}(t)\,dt \tag{6.14}$$

and holds the answer for further processing. The S/H device produces results superior to those obtained when regular op amps are used, since the leakage current at the switch node is exceptionally low. The R_c and C_c elements prevent oscillations during reset, and their product should be at least 0.02 times the integration time constant of $R_I C_I$. (For the simpler integrate-and-reset function without using a hold, an ordinary op amp is substituted for the upper S/H circuit.)

The reset operation is also essential in peak detectors. Many peak detector circuits have been published, but few illustrate practical methods of resetting the circuit. In the scheme of Figure 6.48, the S/H input stage

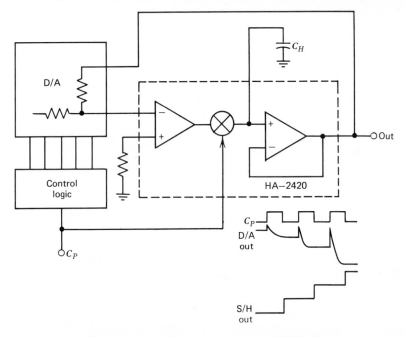

Figure 6.45 Deglitcher circuit using an S/H device.

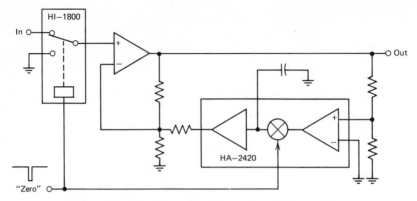

Figure 6.46 Circuit for automatic zeroing of offset voltage.

acts as a gated amplifier for discharging the holding capacitor C_H. The output stage serves as a very-low-bias-current buffer stage. Drift is caused primarily by leakage of the best available diodes, rather than by the buffer or reset stage.

The external op amp must be capable of driving C_H without instability. Another S/H input stage is excellent for this op amp because of its stability and good slew rate when driving capacitive loads.

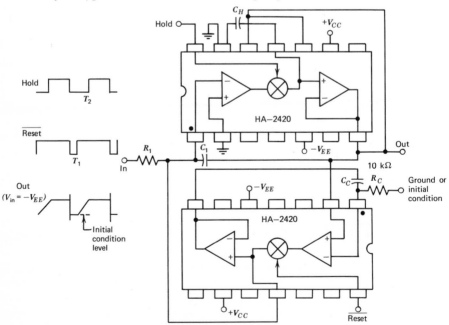

Figure 6.47 Integrator with hold and reset capability.

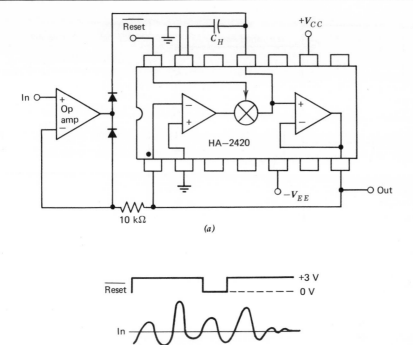

Figure 6.48 Positive peak detector. (*a*) Circuit connections. (*b*) Wave forms.

The two diodes may be reversed to form a negative peak detector. However, the bias current and offset voltage of the S/H output stage produce better results when connected as the positive detector, as in Figure 6.48. A simple inverting amplifier preceding the positive peak detector is recommended for detecting negative peaks.

A final useful application of sample-and-hold techniques is illustrated in Figure 6.49. The diagram shows how fast, repetitive wave forms can be slowed down and plotted using sampling techniques. The input signal is much too fast to be tracked directly by the *X–Y* recorder. However, signal sampling allows the recorder to be driven as slowly as necessary.

To operate the system, the wave form is first synchronized on the oscilloscope. When the potentiometer connected to the recorder is slowly advanced, the wave form will be reproduced. The S/H device samples for a very short interval, once during each horizontal sweep of the scope. The sampling instant is determined by the potentiometer at the instant when

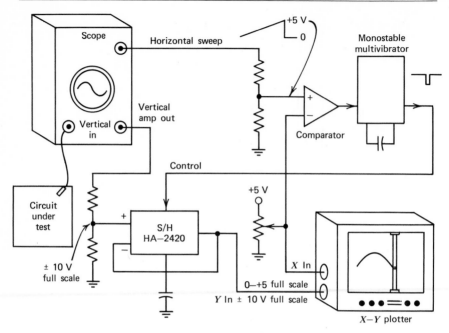

Figure 6.49 Sampling technique for plotting high-speed wave forms.

the horizontal sweep wave form corresponds to the X position of the recorder. The principle illustrated here can be applied to many systems for wave-form analysis.

6.4 CONCLUSIONS

This chapter has described some specialized amplifiers that have evolved from the basic operational amplifiers. We have investigated a class of programmable amplifiers whose internal performance characteristics can be easily changed by external logic and biasing techniques. Cited applications of these programmable amplifiers included analog switches, controlled oscillators, and selectable-gain amplifiers. The chopper-stabilized amplifier was also described as another example of an advanced amplifier. We saw how "choppers" are well suited for applications of self-zeroing instrumentation amplifiers, precision integrators, and high-slew-rate, low drift amplifiers. As a final example we discussed the sample-and-hold gated operational amplifier, which provides voltage gain, analog demultiplexing, and A/D interfacing, in S/H applications. Circuits were also presented for deglitching digital systems, automatic zeroing op amp offset voltage, peak detection, and interfacing an oscilloscope with an X–Y plotter.

BIBLIOGRAPHY

1. Ahmed, H., and P. J. Spreadbury, *Electronics for Engineers, An Introduction,* Cambridge University Press, London, 1973.
2. Angelo, E. James, Jr., *Electronics: BJTs, FETs, and Microcircuits,* McGraw-Hill, New York, 1969.
3. Barna, Arpad, *Operational Amplifiers,* Wiley, New York, 1971.
4. Belove, Charles, Harry Schachter, and Donald L. Schilling, *Digital and Analog Systems, Circuits, and Devices: An Introduction,* McGraw-Hill, New York, 1973.
5. Connelly, J. A., N. C. Currie, and D. S. Bonnet, "Op Amp Has Sixteen-Step Digital Gain Control," *Electronic Design News,* **19,** No. 9 (May 5, 1974), pp. 75–77.
6. Deboo, Gordon J., and Clifford N. Burros, *Integrated Circuits and Semiconductor Devices: Theory and Application,* McGraw-Hill, New York, 1971.
7. Fitchen, Franklin C., *Electronic Integrated Circuits and Systems,* Van Nostrand-Reinhold, New York, 1970.
8. Graeme, Jerald G., Gene E. Tobey, and Lawrence P. Huelsman, *Operational Amplifiers: Design and Applications,* McGraw-Hill, New York, 1971.
9. Graeme, Jerald G., *Applications of Operational Amplifiers: Third Generation Techniques,* McGraw-Hill, New York, 1973.
10. Grebene, Alan B., *Analog Integrated Circuit Design,* Van Nostrand-Reinhold, New York, 1972.
11. Jones, Don, and Robert W. Webb, "Chopper-Stabilized Op Amp Combines MOS and Bipolar Elements on One Chip," *Electronics,* **46,** No. 20 (September 27, 1973), pp. 110–114.
12. Kuo, B. C., *Automatic Control Systems,* Prentice-Hall, Englewood Cliffs, N.J., 1962.
13. Kuo, B. C., *Analysis and Synthesis of Sampled-Data Control Systems,* Prentice-Hall, Englewood Cliffs, N.J., 1963.
14. Melen, Roger, and Harry Garland, *Understanding IC Operational Amplifiers,* Sams, Indianapolis, 1971.
15. Millman, J., and C. C. Halkias, *Integrated Electronics: Analog and Digital Circuits and Systems,* McGraw-Hill, New York, 1972.

CHAPTER 7

ANALOG MULTIPLEXERS

Multiplexing is a popular technique in which data from several signal sources (or channels) are combined in a prescribed orderly fashion onto a single line. Time-division multiplexing can be simply pictured as a rotating commutator that momentarily and sequentially connects each of the several inputs to the single common output. An analog multiplexer, then, is a circuit that serially switches a number of different analog input signals onto a single line or channel. Any individual input channel normally is accessed through a digital address code applied to some digital inputs.

Although monolithic MOS analog multiplexers and switches have been available for several years, they have had several shortcomings. Most important, leakage currents at ambient temperature above +60°C are excessive for most applications. Furthermore, the effective series resistance of the switching elements tends to change drastically with changes in the analog input level. Finally, switching times tend to be excessive, on the order of 1.0 μs.

A recent fabrication process breakthrough, combining CMOS with dielectric isolation, considerably lessens these problems and makes practical many new applications. Typical input leakage current at +125°C with DI CMOS is 1.0 nA with 250 ns switching time.

In this chapter we examine the functional organization of analog multiplexers, define the important static and dynamic parameters, and present several useful multiplexer applications.

7.1 FUNCTIONAL OPERATION OF MULTIPLEXERS

Figure 7.1a is a functional block diagram of a 16-channel CMOS analog multiplexer. Any channel can be accessed by simply applying the proper digital address to the digital inputs as given in Figure 7.1b. The 4-bit

270

address is binary coded, allowing a random access of 2^4, or 16 channels. The additional digital input, identified as the enable input, allows "off–on" control of all 16 channels, which is useful in systems requiring more than one multiplexer.

All five of the digital inputs are compatible with TTL logic levels. However, internal buffering and level shifting on the monolithic chip translates the TTL inputs to the appropriate CMOS voltage levels. Using positive logic and applying all logic zeros (L) to A_0 through A_3 and a logic 1 (H) to the En input, it is easily determined that only the first NAND gate is satisfied. The output at this gate is a logic 0, which is applied directly to the PMOS switch. The NAND gate output is also inverted and simultaneously applied as a logic one to the NMOS switch. As we discussed in

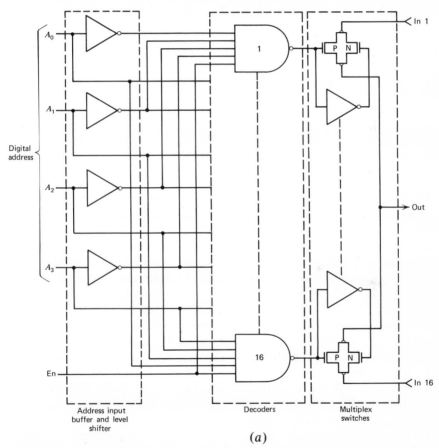

(a)

Figure 7.1   Sixteen-channel analog switch. (a) Functional block diagram. (b) Channel-select code.

A_3	A_2	A_1	A_0	En	"On" channel
X	X	X	X	L	None
L	L	L	L	H	1
L	L	L	H	H	2
L	L	H	L	H	3
L	L	H	H	H	4
L	H	L	L	H	5
L	H	L	H	H	6
L	H	H	L	H	7
L	H	H	H	H	8
H	L	L	L	H	9
H	L	L	H	H	10
H	L	H	L	H	11
H	L	H	H	H	12
H	H	L	L	H	13
H	H	L	H	H	14
H	H	H	L	H	15
H	H	H	H	H	16

(b)

Figure 7.1 (cont.)

Section 3.15, this parallel PMOS–NMOS switch arrangement presents a near-constant "on" resistance between the source and drain (input and output) terminals. Both P- and NMOS devices are turned "on," connecting the input of channel 1 to the common output.

Complete circuit diagrams of the input buffer, level shifter, address decoder, and multiplexer switch appear in Figure 7.2. The address input buffer and level shifter have an internal TTL reference circuit to provide an internal +5 V reference for TTL compatibility. If reference levels in excess of 5 V are necessary, the desired level can be applied directly to the V_{ref} terminal. The digital input gates are protected from electrostatic charges by the 200 Ω resistor and the clamping diodes.

The address decoder circuit of Figure 7.2b is the NAND gate and inverter combination. If all inputs to the serially connected NMOS devices are at $V+$ corresponding to a logic 1, all these NMOS FETs conduct, resulting in an "on" switch for that channel. However, if any one or more of the five digital inputs is at a logic 0, at least one of the NMOS FETs is "off." When this occurs, at least one of the parallel PMOS FETs will be "on," which reverses the polarity of the switch driver voltage.

The multiplex switch of Figure 7.2c utilizes the overvoltage protection

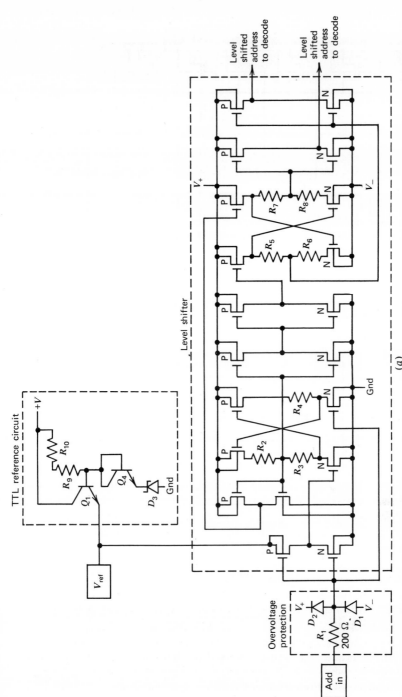

Figure 7.2 Complete circuit diagrams of analog switch. (*a*) Address input buffer and input level shifter. (*b*) Address decoder. (*c*) Multiplexer switch.

273

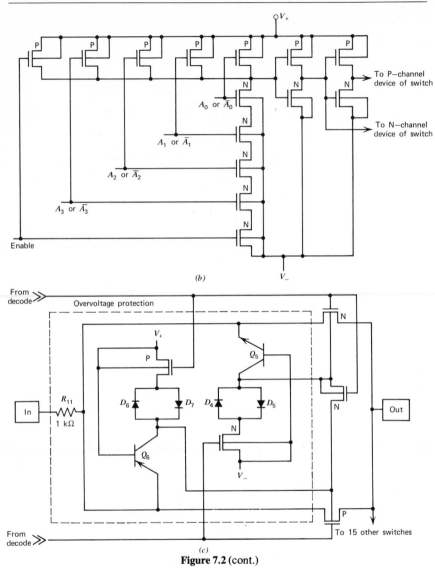

Figure 7.2 (cont.)

circuit discussed in Section 3.15. Under the normal operating voltage range, the switch is simply a 1 kΩ resistor in series with the parallel P- and NMOS FET pair.

With this background in multiplexer operation, let us now consider many of the practical parameters used to characterize the actual device and to distinguish it from its idealistic counterpart. It is these parameters which will impose limitations in various applications.

7.2 MULTIPLEXER PARAMETERS

Parameters describing practical multiplexer operation are best examined when subdivided into three categories:

1. Parameters of the analog signal channel.
2. Parameters of the digital address lines.
3. Switching parameters.

We now examine each of these classifications.

Analog-Signal Channel Parameters

Analog Signal Range ($\pm V_s$) The maximum safe input voltage range and usually specified as being within the $\pm V$ supply voltages. Exceeding $\pm V_s$ can result in channel interaction and/or device destruction if no input overvoltage protection is provided.

"On" Resistance (R_{on}) The maximum resistance associated with an "on" channel over a specified range of input voltages which is usually less than the analog signal range. In most applications the "on" resistance is of secondary importance, since the multiplexer is usually driven into a high-impedance buffer amplifier so that the voltage drop across the switch is approximately zero. When driving into low-impedance loads, however, this voltage drop can become significant and produce intolerable transfer errors.

"Off" Input Leakage Current ($I_{s(off)}$) The maximum current measured at the input of an "off" channel with a specified voltage ($\pm V$) applied to both input and output in any combination. This current is usually in the picoampere range and is due mainly to source–body and surface leakages of the channel. In most applications this small current presents few problems.

"Off" Output Leakage Current ($I_{D(off)}$) The effective sum of all reverse drain–body leakage currents measured at the output. When driving into a high-impedance buffer, the output leakage current will effectively flow through the "on" resistance and the source impedance to develop an error voltage across the switch. The error can be minimized through both low values of source and R_{on} resistances.

"On" Channel Leakage Current ($I_{D(on)}$) The current flowing through the source–body and drain–body diodes of the "on" channel. This current is measured by applying a specified voltage source to input and output

simultaneously and determining the current through the voltage source. In certain worst-case conditions, where $I_{D(off)}$ and $I_{D(on)}$ are in the same direction and consequently add, the error voltage across the "on" channel will be a maximum.

Digital Input Parameters

Input Low Threshold (V_{AL})   The maximum allowable voltage that can be applied to the digital input and still be recognized by the decoder as a logic low input. For TTL and CMOS compatibility, voltages less than 0.8 V are normally specified.

Input High Threshold (V_{AH})   The minimum voltage that can be recognized as a logic high. For TTL and CMOS logic $V_{AH} \geq 2.4$ V and $V_{AH} \geq 6$ V, respectively, are usually specified.

Input High Leakage Current (I_{AH})   The current measured at the digital input with a logic high applied. A large driver-source impedance coupled with a series-limiting resistor for protection can result in a degradation of the voltage at the digital input, causing the "high" state to be unrecognizable.

Input Low Leakage Current (I_{AL})   The current measured at the digital input with a logic low applied. Normally, this current will flow from the digital input and into the logic driver source. When relatively large series-limiting resistors are used to protect the digital inputs, care must be taken to avoid excess voltage drop that would cause V_{AL} to be exceeded.

Switching Parameters

Access Time (T_A)   The total time required to activate an "off" switch to the "on" state. Normally the access time is measured from initiation of the digital input pulse to the 90% point of the output peak-to-peak transition, as in Figure 7.3. The access time is composed of two parts: a driver delay t_d, and an exponential rise t_e. The driver delay is an intentional delay introduced to avoid momentary shorting between signal channels; t_e is due to the output capacitance and the $R_{on} + R_{source}$ resistances. The maximum frequency at which the address inputs can be toggled is directly proportional to this access time. Typical access times for protected and unprotected devices are 500 and 250 ns, respectively. The protected device has a larger "on" resistance, thereby increasing the $R_{on}C_{out}$ time constant. Protected devices also incorporate the break-before-make delay (~80 ns), whereas the unprotected device utilizes the make-before-break overlap (~100 ns).

Make-Before-Break Overlap ($t_{overlap}$)   The time overlap during the transition from one channel to another when both channels are "on." This

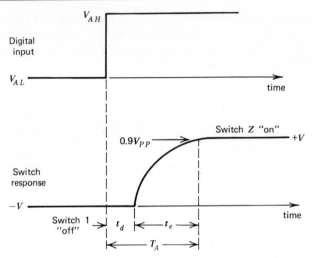

Figure 7.3 Analog switch access time.

scheme effectively permits momentary channel-to-channel shorting but usually results in fast access times. An overlap of 100 ns is typical for a nominal access time of 250 ns.

Break-Before-Make-Delay (t_{open}) The time delay during the transition from one channel to the next when all channels are "off." This requires that the turn-off time delay of the previous switch position be shorter than the turn-on delay of the selected switch position. The obvious advantage of this scheme is that no channel interaction can exist. However, programmed time delays to assure this type of action will result in longer access times of typically 500 ns.

Channel Input Capacitance ($C_{s(off)}$) The capacitance to ground seen looking into an "off" switch. The source–body capacitance is the major contributor and is usually very small (<5 pF).

Channel Output Capacitance ($C_{D(off)}$) The capacitance seen looking into the output of the multiplexer. This is basically the sum of the drain–body capacitances of all the channels. The product of the output capacitance and ($R_{on} + R_{source}$) determines the RC time constant associated with the access time. Thus for maximum speed the source impedance must be as small as possible. The $C_{D(off)}$ is directly proportional to the number of analog channels present on the device; for a 16-channel switch, 50 pF is typical.

Input-to-Output Capacitance ($C_{DS(off)}$) The capacitance between the input and output terminals when the multiplexer is off.

"Off" Isolation   The feedthrough of a signal through an "off" switch to the output. Isolation is specified at a particular frequency and usually expressed in decibels. Major feedthrough in CMOS occurs through the source–body and body–drain capacitances and has a more pronounced effect at higher frequencies. To minimize crosstalk, low values of source and load impedance should be employed.

In summary, we see that the multiplexer is basically a system in itself whose primary function is to switch analog signals. The efficiency at which a signal is transferred from input to output is dependent not only on the characteristics of the device but on how it is interfaced in an actual system. For example, leakage currents can produce a significant error when large source resistances are employed. However, if the analog system designer understands the limitations of the device, he can usually design around these nonideal characteristics.

7.3   MULTIPLEXER APPLICATIONS

Multiplexers are most frequently used to reduce the physical size and weight of a system by reducing the number of components required to process a multitude of analog signals. When used in the normal forward mode, the multiplexer transfers several analog inputs to one common output, each channel occupying a selected time slot. Operation of the multiplexer in the reverse mode (i.e., as a demultiplexer) separates serial data on a single input channel into several outputs, each representing a decoded channel.

The multiplexer has found wide acceptance in airborne systems, where weight is always critical in obtaining peak flight performance. Figure 7.4 is the block diagram of a typical multiplexed airborne telemetry system. This multitiered arrangement will allow several hundred analog signals to be multiplexed using a minimum number of components. Note that the M-channel multiplexer (M_1) selects one of the N multiplexers; thus the theoretical maximum number of analog signals that can be processed is MN, where N is the number of channels associated with each multiplexer (N_1 through N_M). The outputs of multiplexer N_1 through N_M could have been made common to eliminate M_1; however, the effective output capacitance and output leakage currents would increase by approximately the factor M and would result in a large access time with significant offset-voltage errors.

Now let us examine two possible ways of implementing the multiplexer system in Figure 7.4. In the first case the M_1 block is eliminated and 256 analog input channels are multiplexed by setting $M = N = 16$. In this case,

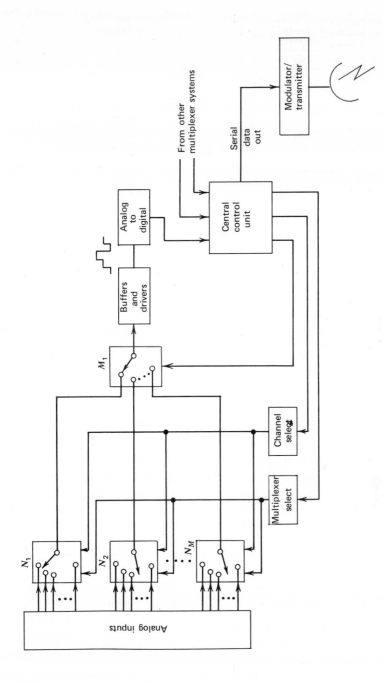

Figure 7.4 Basic airborne telemetry system.

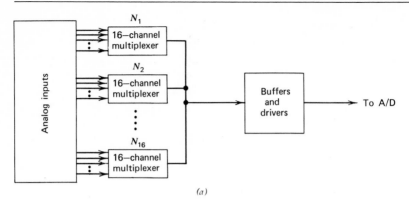

(a)

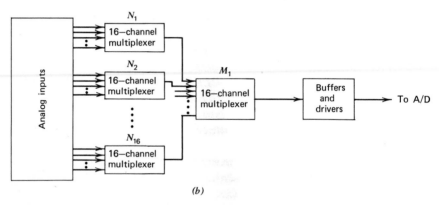

(b)

Figure 7.5 Implementation of telemetry system of Figure 7.4. (a) For Case I without M_1 multiplexer. (b) With an additional M_1 16-channel multiplexer.

all 16 outputs of the N multiplexers are connected together. Our second case is identical to the first except that the 16 outputs of the N multiplexers are now connected to the 16 inputs of the M_1 block. Figure 7.5 illustrates the differences between the two cases. For both cases, we assume that each 16-channel multiplexer has the following worst-case parameters:*

$$R_{on} = 1.5 \text{ k}\Omega \qquad\qquad C_{D(off)} = 50 \text{ pF}$$
$$I_{S(off)} = 50 \text{ nA} \qquad\qquad C_{DS(off)} = 1 \text{ pF}$$
$$I_{D(on)} = 500 \text{ nA} \qquad\qquad I_{D(off)} = 500 \text{ nA}$$
$$T_A = 500 \text{ ns}$$
$$t_{open} = 80 \text{ ns}$$

* These specifications are those for the HI-506A multiplexer.

Case I

An equivalent circuit for the multiplexer system of Figure 7.5a is presented in Figure 7.6. We have arbitrarily selected channel 1 to be the "on" channel and the remaining 255 channels to be "off." Let us now determine the offset-voltage error versus the leakage current.

Although the magnitude and direction of $I_{D(on)}$ is a function of input and output voltage levels, we will assume worst-case magnitude and direction. That is, the amplifier bias current and $I_{D(on)}$ are in the same direction, causing a worst-case offset-voltage error across resistors R_s and R_{on}. Normally R_s and I_{bias} are small enough to be neglected, and the total error voltage is given by

$$-16I_{D(on)}R_{on} = -12 \text{ mV} \tag{7.1}$$

If we require a transfer accuracy of 0.1% at 5 V, Equation 7.1 shows that the maximum tolerable error of 5 mV has been exceeded. Therefore, this leakage current will limit the accuracy that can be obtained.

Next we determine the access time and data rate versus output capacitance for Case I. Access time consists of internal time delays plus

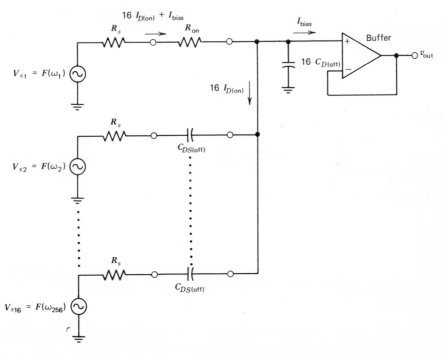

Figure 7.6 Equivalent circuit for Case I of Figure 7.5

the RC time constant associated with the channel. Increasing the output capacitance proportionally increases the RC time constant, causing an increase in the access time. As Figure 7.3 indicates, the access time is measured to 90% of the final level, or $2.3RC$. The access time for the circuit of Figure 7.6 is

$$T_A = 500 \text{ ns} - 2.3R_{on}C_{D(off)} + (2.3)(16)R_{on}C_{D(off)}$$
$$T_A \approx 3\mu s \tag{7.2}$$

Approximately seven time constants are required to achieve 0.1% accuracy. Therefore, the minimum sampling period is given by

$$T_m = 500 \text{ ns} - 2.3R_{on}C_{D(off)} + 7(16)R_{on}C_{D(off)}$$
$$T_m \approx 9 \ \mu s \tag{7.3}$$

The maximum sampling rate is

$$R_m = \frac{1}{T_m} = 111,000 \text{ samples/s} \tag{7.4}$$

The number of analog channels (M), and the maximum channel frequency (f) are related according to

$$2\sum_{N=1}^{M} f_N \leq R_m \tag{7.5}$$

where N is the channel number.

The number of samples per second that must be taken for each channel is a function of the maximum channel frequency. According to the Nyquist sampling theorem, taking at least two samples per period of the maximum frequency component present will ensure no loss of transmitted information. Applying the sampling theorem to this case, where there are 256 channels and $R_m = 111,000$ samples/s, the maximum channel frequency that can be processed with all channels having the same frequency bound is

$$(2)(256)(f_{max}) = 111,000$$
$$f_{max} = 216 \text{ Hz} \tag{7.6}$$

This represents an average channel frequency for the system resulting in an average channel rate (CR) of

$$CR = 432 \text{ samples/s} \tag{7.7}$$

Naturally, not all channels would be sampled at the same rate, since some analog signals represent slowly varying inputs (e.g., tank pressure and volume), whereas other inputs are from rapidly varying sources (e.g.,

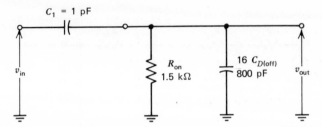

Figure 7.7 Equivalent circuit for Case I for determination of channel-to-channel "crosstalk."

acoustical and acceleration transducers) which require very high sampling rates.

To determine the channel-to-channel "crosstalk" for the Case I system, we examine an equivalent circuit for this determination (Figure 7.7), which represents an "off" channel. The transfer function for this circuit is

$$\frac{V_{out}(s)}{V_{in}(s)} = \frac{sR_{on}C_1}{1 + sR_{on}(C_0 + C_1)} \tag{7.8}$$

where $C_1 = C_{DS(off)}$ and $C_0 = 16C_{D(off)}$. The "off isolation" for the 16-channel multiplexer at 500 kHz is 65 dB. The isolation in decibels at 500 kHz is found from Equation 7.8 to be

$$\text{"off isolation"} = 20 \log \left| \frac{V_{out}}{V_{in}} \right| = 58 \text{ dB} \tag{7.9}$$

Recalling that the maximum channel frequency cannot be greater than 50 kHz (as seen from the maximum sampling rate), the "off isolation" at 50 kHz is 67 dB.

Case II

Now let us examine the use of an additional multiplexer (see Figure 7.5b). An equivalent circuit for this case is diagrammed in Figure 7.8a. The effective offset-voltage error contribution of this scheme is only 3 times that of a single multiplexer system. The circuit for finding the error voltage appears in Figure 7.8b and can be found as

$$-3R_{on}I_{D(on)} = -2.3 \text{ mV} \tag{7.10}$$

Here we have assumed $R_s \ll R_{on}$.

Using the equivalent circuit of Figure 7.8c to determine the access time gives a transfer function of

$$\frac{V_{out}(s)}{V_{in}(s)} = \frac{1}{(sR_{on}C_{D(off)}^2 + 3sR_{on}C_{D(off)} + 1)} = \frac{1}{(T_1s + 1)(T_2s + 1)} \tag{7.11}$$

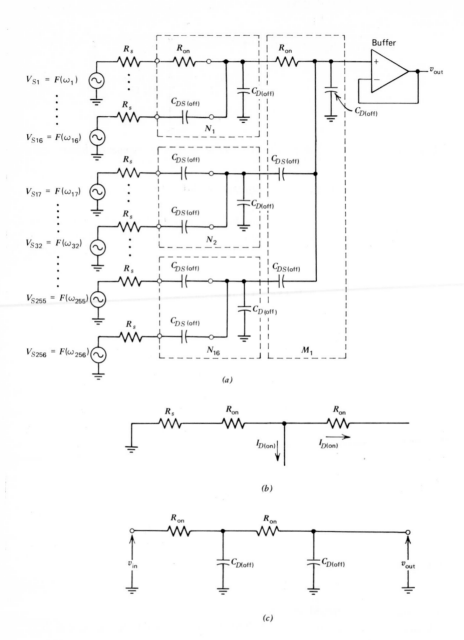

Figure 7.8 Equivalent circuits for Case II. (*a*) Circuit for Figure 7.5*b*. (*b*) Circuit for finding offset-voltage error versus leakage current. (*c*) Circuit for determination of access time and data rate.

where

$$T_1 = 2.61 R_{on} C_{D(off)} \qquad (7.12)$$

and

$$T_2 = 0.38 R_{on} C_{D(off)} \qquad (7.13)$$

By transformation to the time domain with a step-function input, the output voltage as a function of time becomes

$$v_{out}(t) = 1 - 1.17 e^{-t/T_1} + 0.17 e^{-t/T_2} \qquad (7.14)$$

The time required for the output voltage given by Equation 7.14 to reach 90% of the final value is $6.5 R_{on} C_{D(off)}$. Therefore, the access time can be found as

$$T_A = 500 \text{ ns} - 2.3 R_{on} C_{D(off)} + 6.5 R_{on} C_{D(off)}$$
$$T_A = 815 \text{ ns} \qquad (7.15)$$

A time of approximately $18 R_{on} C_{D(off)}$ is required to achieve a 0.1% accuracy. Therefore, the minimum sampling period is found from

$$T_m = 500 \text{ ns} - 2.3 R_{on} C_{D(off)} + 18 R_{on} C_{D(off)}$$
$$T_m = 1.7 \text{ } \mu s \qquad (7.16)$$

From this equation, the maximum sampling rate is 590,000 samples/s (SPS) with an average channel rate of 2300 SPS. The "off isolation" of a single multiplexer (65 dB at 500 kHz) is unaffected using this scheme. Summarizing and comparing the results obtained for these two multiplexer configurations, we have:

Parameter	Case I	Case II
Offset-voltage error	12 mV	2.3 mV
Access time, T_A	3000 ns	815 ns
Maximum sampling rate, R_m	111,000 SPS	590,000 SPS
Average channel rate, CR	432 SPS	2300 SPS
Off isolation (500 kHz)	58 dB	65 dB

As this summary shows, Case II results in a much superior system. Performance advantages of 6:1 are realized by adding one additional multiplexer, which represents only a small fraction of the total system cost. This example involves only one instance in which knowledge of multiplexer parameters permits the designer to greatly improve the system's performance.

The bilateral property of CMOS multiplexers allows us to use them as demultiplexers. In many cases, point-to-point data transmission requires a

286

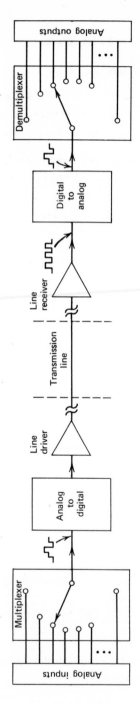

Figure 7.9 Data transmission system.

large number of analog signals to be transmitted over very long distances. The multiplexer/demultiplexer scheme in Figure 7.9 can help to reduce system costs by greatly lessening the number of point-to-point cable runs as well as the number of individual components. At the transmitting end, a multiplexer system is used to combine all analog inputs into distinct time slots. At this point, the analog data could be sent down the transmission line, but line attenuation would severely distort the information. Instead, the analog data are converted by the A/D converter into digital data and transmitted relatively undisturbed by attenuation. At the receiving end, digital data are received in serial form by the line receiver, converted to analog serial data, and applied to the demultiplexer for a serial-to-parallel conversion. For maximum efficiency, the multitiered scheme for multiplexing and demultiplexing should be used as before.

Multiplexers are currently being used extensively for digital signal processing. As an example, consider a mechanical series-switched digital filter (Figure 7.10) in which both the input and output signals are sampled by a pair of rotary switches revolving at a frequency f_o. Commutation at f_o will successively expose each RC filter section to the input signal for one-fourth the period T_o. The charging rate for each capacitor is limited by the RC time constant, which is made large compared to the time that the switch is closed. Therefore, several samples are required for each capacitor to charge to the average value of the input voltage segment. If the input frequency is the same as the commutation frequency f_o, each

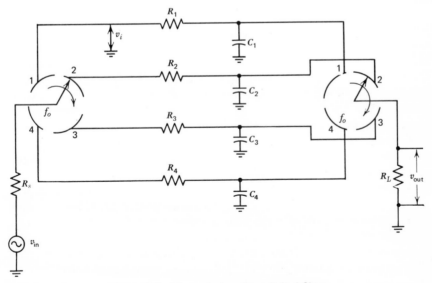

Figure 7.10 Mechanical series-switched filter.

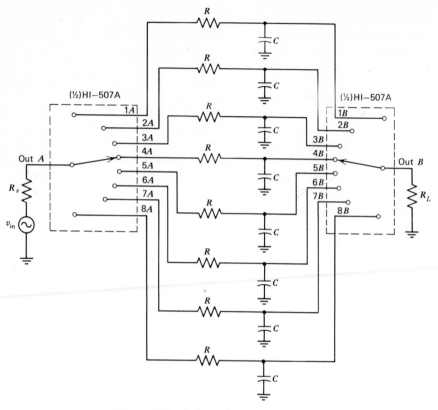

Figure 7.11 Series-switched digital filter.

time a switch pair is closed, one-fourth of the sine wave is impressed on the RC filter. However, if the input frequency differs from f_o, the average value stored on the capacitor will be different for each successive time interval. The rate at which the average value will vary is determined by the difference between f_o and the input frequency. Therefore, for large differences, the average value stored on the capacitor will be near zero. Thus the low-pass filter characteristics are translated into a bandpass filter centered about the commutation frequency f_o.

An 8-channel CMOS multiplexer can be used to replace the mechanical rotary switches (see Figure 7.11). This multiplexer provides constant "on" resistance over a wide analog range, minimizing the imbalance between RC sections. This in turn will minimize the third-order image response that results from imbalance. Furthermore, the uniformity of "on" resistance among channels will reduce "zero-signal noise" and in most cases will eliminate cumbersome in-line balance potentiometers.

The "zero-signal noise" is measured at the output with zero input voltage. This parameter serves to indicate the maximum attenuation possible for frequencies out of the filter's passband. The maximum attenuation is usually given in decibels and is calculated from

$$\text{max. attenuation} = 20 \log_{10}\left(\frac{V_{in}}{V_{zero}}\right) \qquad (7.17)$$

where V_{in} is the input-signal voltage and V_{zero} is the zero-signal voltage. The maximum attenuation will be defined by the zero-signal voltage when low-pass filters having greater than 6 dB/octave rolloff are used. The zero-signal voltage, if large enough, will significantly affect the filter's bandwidth, filter Q, and the shape of the bandpass. To see how other parameters affect the bandpass response, let us consider the options available for increasing the circuit Q at a given resonant frequency f_o.

1. *Increase the Number (N) of RC Sections*   Maintaining a given f_o requires an increase in the clock frequency f_c according to

$$f_c = Nf_o \qquad (7.18)$$

Increasing N reduces the sampling interval for each section, thereby allowing less time for the out-of-band frequencies to cause a change in capacitor voltage. The maximum value of N is limited by the maximum toggle frequency of the multiplexer.

2. *Lower the Clock Duty Cycle*   The clock duty cycle can be easily reduced by driving the multiplexer with a monostable multivibrator triggered by the clock frequency. This scheme has the same effect as increasing N but is independent of the number of sections. However, as the ratio of switch "on" time to access time decreases to a practical value less than about 60, the zero-signal voltage produces a significant change in the filter's rolloff characteristics, with a corresponding degradation of Q and the maximum attainable attenuation. Therefore, a tradeoff exists between the maximum Q and the zero-signal voltage.

3. *Increase the Filter RC Time Constant*   If the number of sections as well as the clock frequency are fixed, then increasing the RC time constant reduces the bandwidth, which increases the Q. The multiplexer leakage current and crosstalk will limit the maximum RC that can be practically used.

Another type of digital filter having a performance similar to that of the series-switched configuration just considered is the shunt-switched filter. Figure 7.12 shows the practical implementation of this digital filter using a 16-channel multiplexer. The multiplexer successively grounds a series of

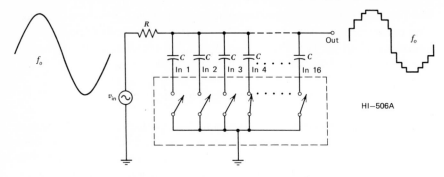

Figure 7.12 Shunt-switched digital filter.

capacitors. The device will accommodate up to 16 RC, sections resulting in very high Q circuits and almost nonexistent imbalance effects.

7.4 CONCLUSIONS

In this chapter we have discussed the operation and performance parameters of analog multiplexers. We have seen how these parameters can be related to the operation of practical multiplexer systems. Multiplexer applications were illustrated for data transmission systems, series-switched filters, and shunt-switched filters.

BIBLIOGRAPHY

1. Eimbinder, Jerry, *Semiconductor Memories*, Wiley, New York, 1971.
2. Graeme, Jerald G., Gene E. Tobey, and Lawrence P. Huelsman, *Operational Amplifiers: Design and Applications*, McGraw-Hill, New York, 1971.
3. Graeme, Jerald G., *Applications of Operational Amplifiers: Third Generation Techniques*, McGraw-Hill, New York, 1973.
4. Grebene, Alan B. *Analog Integrated Circuit Design*, Van Nostrand-Reinhold, New York, 1972.
5. Kuo, B. C., *Automatic Control Systems*, Prentice-Hall, Englewood Cliffs, N.J., 1962.
6. Kuo, B.C., *Analysis and Synthesis of Sampled-Data Control Systems*, Prentice-Hall, Englewood Cliffs, N.J., 1963.

CHAPTER 8

DIGITAL–TO–ANALOG AND ANALOG–TO–DIGITAL INTERFACING

Many systems handle both linear and digital data—in fact, it would be very difficult to describe a system that is either purely linear or purely digital. Even simple linear instruments usually contain range-switching functions that can be considered digital. Virtually every digital system requires one or more power-supply regulators, which are linear devices. Because it is difficult to completely separate linear and digital functions, the system designer and/or system analyst must be familiar with both types of device.

Many operations can be implemented using either linear or digital circuit techniques. The system designer, to optimize cost and performance must carefully partition his system, weighing the advantages of currently available large-scale digital circuits against the circuit simplicity offered by the latest low-cost, highly accurate and versatile linear devices.

In this chapter we describe a number of integrated circuits developed recently to interface between linear and digital signals. We also illustrate the operational concepts that must be considered when these devices are implemented in systems designed to accommodate specific applications.

8.1 DIGITAL-TO-ANALOG CONVERSION

In systems employing digital–to–analog (D/A) conversion, discrete bits of information in the form of logic 1s and logic 0s represent the input signal, which is then converted to an analog signal whose magnitude is linearly

291

related to the value of some physical quantity. D/A converters are most often used to interface between some digital data processor, such as a digital computer, and an output device having linear characteristics. Some of the more common specific applications of D/A converters include:

1. Data processing output interface; driver for displays, plotters, and other devices.
2. Programmable power supply or function generator in automatic test equipment.
3. Tool interface in numerical controlled machining.
4. Interface for automatic process control to control temperature, flow rates, and other variables.
5. Digital communications: digital-to-audio interface.
6. Feedback network in A/D converters.

D/A Terminology

Before we examine specific conversion approaches and applications, let us define some of the common terms and parameters encountered in D/A conversion.

Resolution   An indication of the number of possible analog output levels, usually expressed as the number of input bits that the converter will handle. For example, an 8-bit binary-weighted converter will have $2^8 = 256$ possible output levels (including zero). Resolution should not be confused with accuracy, which is sometimes also expressed as a number of bits.

Accuracy   A measure of the deviation of the analog output level from its predicted value under any input combination. Accuracy can be expressed as a percentage of full scale, a number of bits (N bits accuracy $= 1/2^N$ possible error), or a fraction of the least significant bit (LSB). If a converter with M bits resolution has 1/2 LSB accuracy, the possible error is $1/2 \times 1/2^M$. Accuracy may be of the same, higher, or lower order of magnitude as the resolution. The importance of accuracy versus resolution depends on the application. Possible errors in individual bit weights may be cumulative with combinations of bits. Errors in the summation of individual bit weights and changes in these due to temperature variations will degrade the system accuracy. In an effort to make the unit look better on paper, accuracy errors are sometimes separated into linearity, zero drift, full-scale temperature coefficient, and so on. However, unless the system contains provisions for frequent automatic calibration, absolute accuracy, which is the deviation from a theoretically "perfect" D/A output at any input combination, is the only meaningful specification.

Least Significant Bit (LSB)   The digital input bit carrying the lowest numerical weight, or the analog level shift associated with this bit, which is the smallest possible analog step.

Most Significant Bit (MSB)   The digital input bit carrying the highest numerical weight, or the analog level shift associated with this bit. In a binary-weighted converter the MSB creates half the full-scale level shift.

Settling Time   The total time measured from a digital input change to the time the analog output reaches its new value within a specified error band. The transition from one level to another is not always smooth; spikes and ringing may occur.

Basic D/A Conversion Techniques

Figure 8.1 is a block-diagram representation of a basic D/A converter. The converter accepts a 3-bit plus-sign digital input and converts it into an equivalent analog voltage. The essential parts of the converter are the

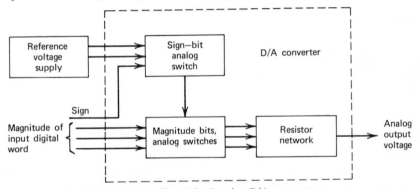

Figure 8.1   Three-bit plus sign D/A converter.

reference-voltage supply, the D/A decoder containing the magnitude bits and analog switches, and the resistor network.

Each digital input (except the sign bit) drives a separate analog switch that either connects or leaves open, depending on the input logic state, the reference supply voltage to the proper terminal of the resistor network. The sign-bit switch is used for discrimination of positive and negative numbers.

Digital Controller

Figure 8.2 illustrates a common D/A system where the desired output is a shaft position of a dc motor. The various combinations correspond to a particular shaft position of the motor, *M*. The job of the D/A converter is

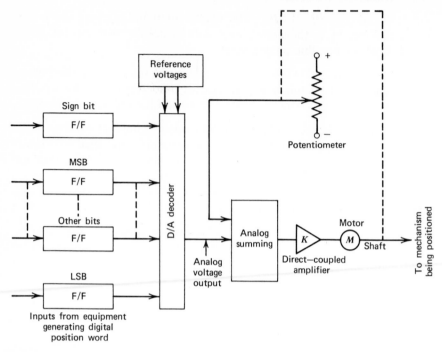

Figure 8.2 Digital controller using D/A converter. (Reprinted from *Analog-to-Digital, Digital-to-Analog Conversion Techniques* by David Hoeschele, Jr. © 1968 by John Wiley & Sons.)

to decode the digital input word into an analog voltage, which is then used as one input to drive the motor turning the shaft to the desired position.

Negative feedback is employed in the digital controller by the summing network and potentiometer. The analog voltage from the converter is summed in a summing network with the output of a potentiometer whose shaft is connected to the shaft of the motor. The output from the analog summer is an error signal that is amplified before driving the motor in a direction to decrease the error signal. When the motor shaft is in the correct position in accordance with the input digital word, the analog output of the potentiometer will equal the analog output of the D/A converter. Thus the error signal is zero, and there is no drive signal for the motor.

An alternative scheme for a digital controller is given in Figure 8.3. This system differs from the controller in Figure 8.2 in that the feedback is obtained from a code wheel driven by the motor. The code wheel generates a digital word that is a function of the present shaft position. The output of the code wheel is digitally compared with the actual

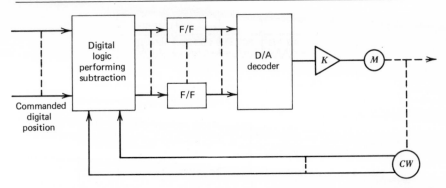

Figure 8.3 Digital controller using digital summation. (Reprinted from *Analog-to-Digital, Digital-to-Analog Conversion Techniques* by David Hoeschele, Jr. © 1968 by John Wiley & Sons.)

commanded position of the input digital word. The comparison is really a subtraction that produces an error signal in digital form to be converted in the D/A decoder. The analog output from the D/A drives the motor in a direction to decrease the error and at a speed proportional to the error.

D/A Devices

One type of monolithic D/A converter (Figure 8.4) incorporates a thin-film resistor ladder network on the same monolithic chip as the analog switches. Along with the advantages of small size and monolithic reliability, this feature allows higher operational speed and faster settling times than are typically obtained with discrete or hybrid converters. It is also superior to a system of separate switches and ladder networks in that overall performance is guaranteed and there is no need to add the possible errors of separate components.

In the functional diagram, an external reference supply, usually +5.0 V, is connected between the +ref and the −ref pins. The output levels will be directly proportional to the differential reference voltage. Variations of the other +5 V and −15 V power supplies have negligible effect on the analog output. Either the +ref or the −ref pin may be grounded; thus reference supplies of either polarity may be accommodated. The smaller value resistors between +ref and T_1 or between T_2 and T_3 may be externally shorted out, or an external trimmer substituted, for fine adjustment of the full-scale output level. The positive side of the reference supply may be connected to T_2 or T_3 for a 10 V nominal output swing, which is useful in bipolar operation.

Although some monolithic converters contain an internal reference regulator, the temperature coefficient of the devices is consistent with

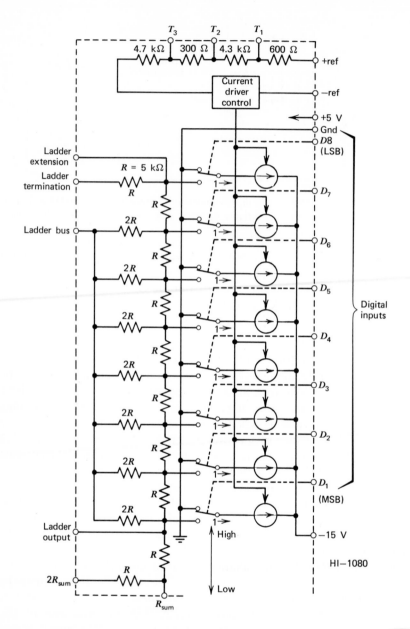

Figure 8.4 Functional diagram of monolithic D/A converter.

only 6 or 7 bits accuracy. For this reason, an external reference supply is utilized in the HI-1080. The reference level is conditioned through the current-driver control where the output current of each of the eight current drivers is determined.

The choice of current sources instead of voltage sources to drive the ladder network presents several advantages. Within the constant-current range of the sources, the current—hence the differential ladder output voltage—will remain constant regardless of variations in the negative supply voltage and the voltage of the ladder return bus. The ladder bus may be returned to a voltage other than ground if desired for offset or bipolar outputs, without affecting its output. The digital and analog grounds can be effectively separated for noise-free operation. Also, switching of a current source rather than a voltage source generally is faster, creates less ringing at the output, and produces smaller power-supply transients.

The digital inputs effectively switch the current-source outputs either to the ladder network or to ground. The inputs are fully compatible with any standard 5 V DTL or TTL logic circuits. A "high" input ($>+2$ V) switches the current source to ground; a "low" input ($<+0.8$ V) switches the current source to the ladder, creating a more negative output voltage.

The ladder network is constructed from high-stability metal-film resistors deposited on the same silicon chip. Identical material is used for the resistors in the reference-supply network and in the current-source circuitry to achieve good temperature stability. An "$R–2R$" ladder network of the type shown in Figure 3.40 is used rather than a binary-weighted resistor network (Figure 3.39) because identical resistors will match better in value and temperature coefficient than nonidentical resistors. Extra resistors provided in the R_{sum} and $2R_{sum}$ terminals are very useful for feedback or summing with external amplifiers or comparators, since these resistors will track almost perfectly with the ladder source resistance. Provision is made at the other end of the ladder for cascading converters for higher resolution. A block diagram of the D/A converter and complete circuit diagrams for the functional block diagram parts appear in Figures 8.5 and 8.6, respectively.

Some of the possible operating modes of the HI-1080 are illustrated in Figure 8.7. Since both the reference-supply terminals and the ladder-bus terminal can be connected to any voltage within ±5 V with respect to power-supply ground, a number of output polarity modes can be achieved. In all cases the ladder output will become more negative with respect to the ladder bus as a digital input is changed from the high to the low state.

In the unipolar–zero reference mode, the ladder bus is grounded. Using

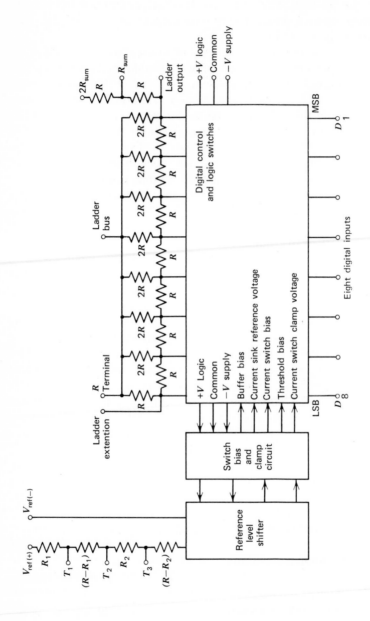

Figure 8.5 Block diagram of 8-bit D/A converter.

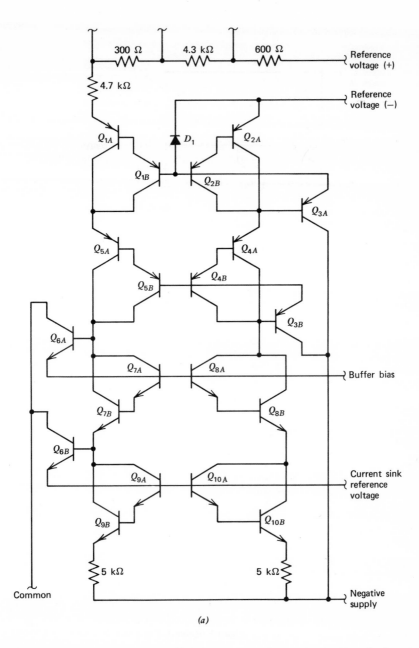

Figure 8.6 Circuit diagrams of 8-bit D/A converter. (*a*) Reference level shifter. (*b*) Current switch bias and clamp. (*c*) Ladder current switch.

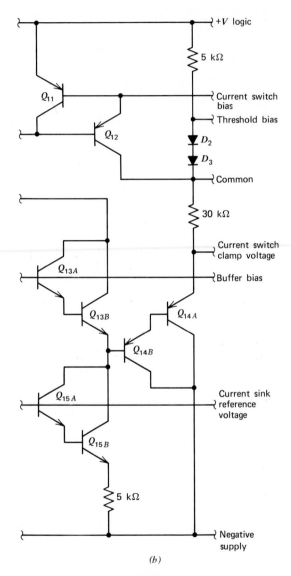

(b)

Figure 8.6 (cont.)

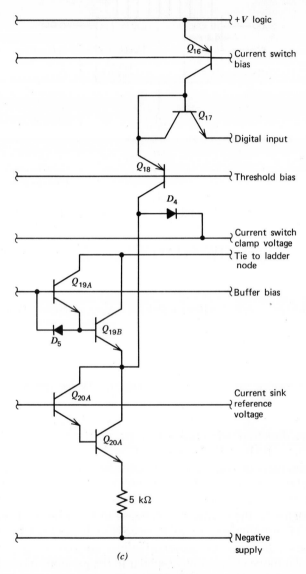

+V logic

Q_{16} Current switch bias

Q_{17} Digital input

Q_{18} Threshold bias

D_4

Current switch clamp voltage

Tie to ladder node

Q_{19A} Buffer bias

D_5 Q_{19B}

Q_{20A} Current sink reference voltage

Q_{20A}

5 kΩ

Negative supply

(c)

Figure 8.6 (cont.)

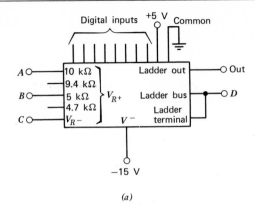

(a)

| | Output range inputs | Connections | | | |
Mode	all high to all low	A	B	C	D
Unipolar zero reference	0 to $-[V_{R^+}-1\text{ LSB}]$	V_{R^+}	NC	GND	GND
Unipolar zero FS	$+\lvert V_{R^+}\rvert$ to $[0+1\text{ LSB}]$	V_{R^+}	NC	GND	V_{R^+}
Bipolar	$\lvert V_{R^+}\rvert$ to $[-V_{R^+}+1\text{ LSB}]$	NC	V_{R^+}	GND	V_{R^+}

(b)

Figure 8.7 Operating modes of D/A converter. (a) Device terminal locations. (b) Terminal connections.

negative logic convention (high = 0, low = 1), the output will increase in the negative direction with increasing input binary number. Either side of the reference supply may be grounded.

In the unipolar–zero full-scale model, the ladder bus is connected to the positive reference voltage; thus the output will be always positive with respect to the reference ground. Now, using positive logic convention (low = 0, high = 1), the output will increase in the positive direction with increasing binary number. It may be necessary to connect V_{R^+} to a lower tap in series with a potentiometer to adjust the zero level.

The bipolar mode connection is similar to the previous mode except the V_{R^+} is connected to T_2 (or T_3 through a potentiometer), and therefore the full-scale excursion is now 10 V. With all inputs low, the output will be most negative (about −4.96 V). With only the MSB high, the output will be zero volts. With all inputs high, the output will be at V_{R^+}.

Figure 8.8 illustrates connections from the converter to an operational amplifier. The inverting connection uses the summing resistors provided on the chip for amplifier feedback. Since the ladder impedance is nominally 5 kΩ, connection of the output to R_{sum} will result in a gain of -1. Connection to $2R_{sum}$ will result in a gain of -2. Any of the operating modes discussed previously may be used.

For a noninverting output the operational amplifier is wired in the conventional manner. The R_{sum} or $2R_{sum}$ resistor could be used to sum an external analog signal of opposite polarity at the amplifier input.

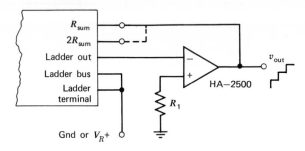

Full–scale Output	Output Feedback Connected to:	R_1
+4.98 V	R_{sum}	2.5 kΩ
+9.96 V	$2R_{sum}$	3.3 kΩ

(a)

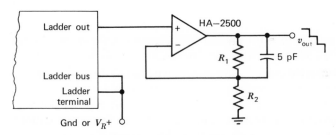

Output range: same as shown in 'operating mode' chart multiplied by $(R_1 + R_2)/R_2$

(b)

Figure 8.8 D/A converter with buffer amplifier connection. (a) Inverting output (more positive with increasing complement of input number). (b) Noninverting output (more negative with increasing complement of input number).

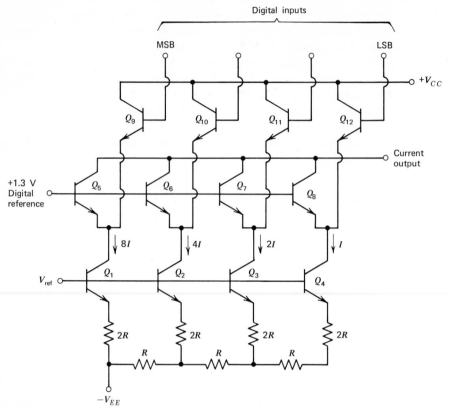

Figure 8.9 Circuit schematic of 4-bit digital-to-analog converter using current-switching technique.

A somewhat different conversion scheme that has been employed in several monolithic D/A converters is shown in Figure 8.9. In this circuit, current always flows through the nodes of the ladder network. These binary-weighted currents are either switched to the output line or diverted to a power line, depending on the digital input.

The ladder network may be either the $R-2R$ type as shown, or single, binary-weighted resistors connected to the negative supply voltage. Transistors Q_1 through Q_4 regulate the currents from the ladder nodes to relative values of $8I$, $4I$, $2I$, and I. The reference voltage at the bases of Q_1 through Q_4 is normally compensated by another transistor on the chip to eliminate variations in I due to changes in V_{BE} and current gain. Further compensation is achieved by designing these four transistors with relative emitter areas of 8, 4, 2, and 1, respectively.

Transistors Q_5 through Q_{12} form four different current switches that switch each of the binary-weighted currents either to the output current line or to the positive power supply, depending on the digital input. Such monolithic current switches are usually fabricated with four switches on a chip, and the chips are graded according to the accuracy of their current ratios. Then, to form an 8-, 10-, or 12-bit D/A converter, the most accurate chip is used for the four most significant bits, since the most accurate current weights are required here.

In the complete circuit schematic of a 6-bit monolithic D/A converter (Figure 8.10a), four functional blocks composing the converter are identified by dashed lines. Figure 8.10b is a block-diagram representation of the converter.

Monolithic 10-bit D/A converters are available which use the current-switching scheme. One such converter (see Figure 8.11) is complete with a reference-voltage supply and an output buffer amplifier on the same chip. The internal reference voltage is buffered by another internal op amp and is used to set a voltage regulator for driving the bases of each binary current source. The current–source emitter voltages are all equal, allowing the use of the R–$2R$ ladder network. These resistors are matched to better than 0.05% over the specified temperature range.

As indicated, this monolithic 10-bit converter uses the current-switching logic in the collectors of the current–source transistors to steer the bit current. A high input (positive logic) will cause each bit current to be drawn from the current-summing line. A low input causes current to be drawn from ground instead of from the sum line. The converter also features a sign bit to produce either a positive or negative analog output voltage. Depending on the state of the sign bit, the sum line current is either presented directly to the summing amplifier or is previously inverted by a current-inverting amplifier. This polarity-controlled current is then converted into a voltage by an internally compensated inverting op amp and feedback resistor, which is closely matched to the reference resistor, R_R.

Cascading D/A Converters

Two 8-bit D/A converters may be cascaded to achieve resolutions from 9 to 15 bits, using the ladder extension terminals, as illustrated in Figure 8.12. Note that input D_8 of the higher significant bit unit is not used. This is necessary to join the two ladders correctly. A person might ask, "Why would anyone want a 12-bit converter with only $8\frac{1}{2}$-bit accuracy?" The answer is that most applications require accuracy expressed as a percentage of actual output rather than as a percentage of full-scale output. One feature of the R–$2R$ ladder network is that errors in terms of millivolt

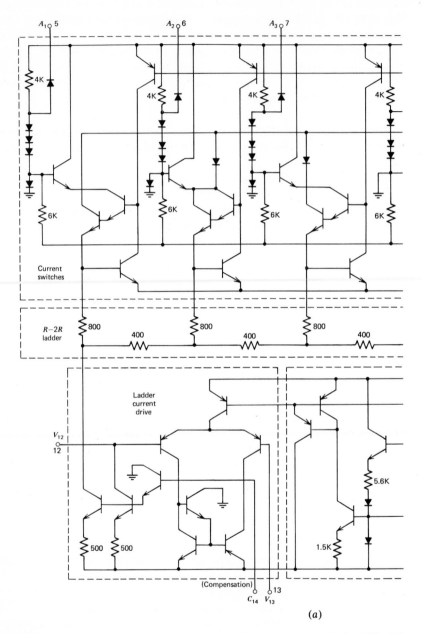

Figure 8.10 Monolithic, 6-bit, multiplying digital-to-analog converter (MC1406L). (*a*) Motorola, Inc.)

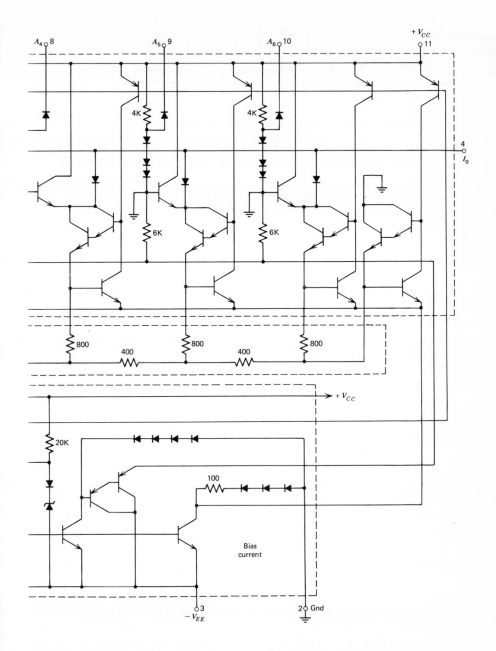

Complete circuit diagram. (b) Block diagram of D/A converter. (Reprinted courtesy of

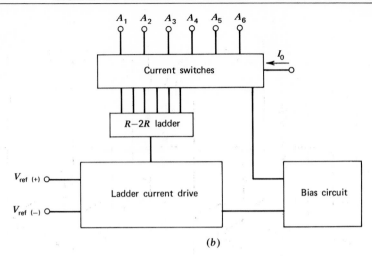

(b)

Figure 8.10 (cont.)

deviation from the predicted output tend to become smaller as the lesser significant bits only are exercised. Thus for the 12-bit converter shown with outputs between 1.25 and 5 V, the errors may be on the order of ±10 mV; but for outputs between 0 and 20 mV the errors will tend to be less than 0.7 mV.

8.2 D/A APPLICATIONS

One of the simplest applications of a D/A converter is a thumbwheel-switch encoder. Here the digital input is obtained manually from a series of ganged switches, as in Figure 8.13. Such a connection requires one set of four ganged switches per decade input. A D/A converter with binary-coded decimal (BCD) coding should be used for accepting the 4-bit per decade input. The switch positions that are logic 0s are connected to ground, and those that are logic 1s are either left open (for TTL) or connected to $+V_{CC}$. Additional sets of ganged switches are added for each decade number to be decoded.

A more sophisticated D/A application utilizes a converter to digitally control a current source. Loads for which current control is important include cathode ray tube deflection coils, motor windings, and pen drives for strip-chart recorders. Figure 8.14 illustrates one possible way of using the output of a D/A converter to control the current in a load device that has one terminal grounded. The op amp A_1 sums the analog output voltage from the D/A converter with the load voltage v_L which has been amplified

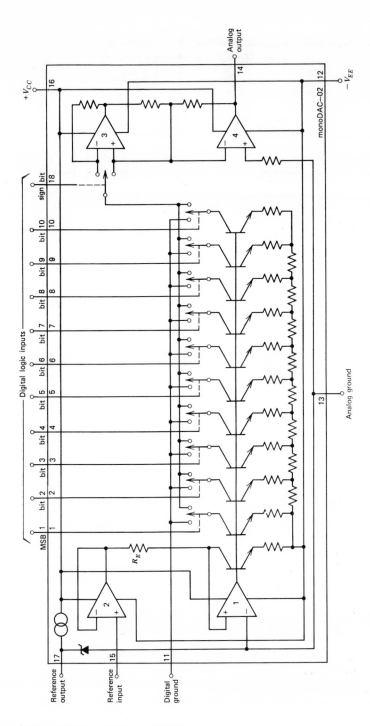

Figure 8.11 Simplified schematic of 10-bit digital-to-analog converter. (From application note, "Applying a Monolithic 10-Bit D/A Converter," Precision Monolithics mono DAC-02, by D. J. Dooley. Reprinted courtesy of Precision Monolithics, Inc.)

309

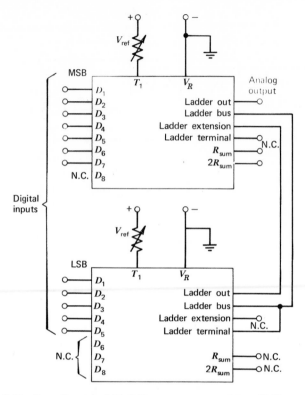

Figure 8.12 Cascading two 8-bit D/A converters to achieve 12-bit resolution.

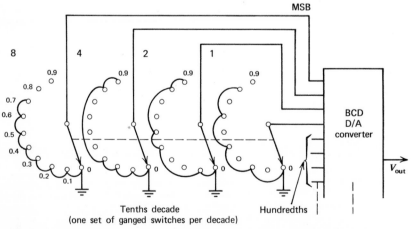

Figure 8.13 Thumbwheel-switch encoder. (Reprinted from *Analog–Digital Conversion Handbook*, with permission of Analog Devices, Inc., Norwood, Mass.)

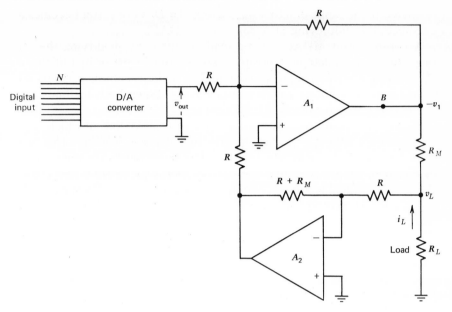

Figure 8.14 Digitally controlled current source. (Reprinted from *Analog–Digital Conversion Handbook*, with permission of Analog Devices, Inc., Norwood, Mass.)

by A_2. Amplifier A_1 sets its output voltage to $-v_1$ such that

$$\frac{v_{\text{out}}}{R} - \frac{v_1}{R} + \frac{v_1 - i_L R_M}{R} = 0 \tag{8.1}$$

Thus we see that

$$i_L = \frac{v_{\text{out}}}{R_M} \tag{8.2}$$

Of course in the general case, the R and R_M resistors can be adjusted for proper scaling. Also if more current is required than can be supplied by the op amp A_1, a current booster could be inserted at point B.

Another important application of D/A converters involves multiplication of two input signals: analog voltage and a digital word. The analog output from a D/A converter can be the product of the two inputs by letting the reference voltage V_R become the analog input. The multiplication operation is easily recognized when the output from a unipolar n-bit binary D/A is represented as

$$v_{\text{out}} = V_R(a_1 2^{-1} + a_2 2^{-2} + a_3 2^{-3} + \cdots + a_n 2^{-n}) \tag{8.3}$$

where V_R is the analog reference voltage and the a_1 through a_n are zero if a bit is a logic 0 and one if a bit is a logic 1. If V_R is an analog signal restricted to the range of 0 to $V_{R(max)}$, a unipolar D/A converter becomes a one-quadrant multiplying D/A converter (MDAC).

A two-quadrant MDAC can be realized either by modifying the V_R range to accommodate both positive and negative inputs or by utilizing a D/A having a sign-bit input provision. The method chosen depends on which input is unipolar and which is bipolar. When both methods are combined, a four-quadrant MDAC results.

Figure 8.15 illustrates a useful application of the MDAC in which two, four-quadrant MDACs can accomplish digital phase shifting. The two analog inputs are $V \sin \omega t$ and $V \cos \omega t$. The digital inputs are logic 1s and 0s coded in such a way that they are proportional to $\sin \theta$ and $\cos \theta$. [Usually the digital inputs are obtained from a programmed memory such as a read-only memory (ROM).] The op amp sums the outputs from the two MDACs to give

$$v_{out} = V \sin (\omega t + \theta) \tag{8.4}$$

Thus the phase-angle delay θ is easily set by the converters' digital inputs. By connecting the V_R inputs together and applying $V \sin \omega t$, the output becomes

$$v_{out} = V \sin \omega t (\sin \theta + \cos \theta) \tag{8.5}$$

Thus this modified connection produces an angular resolver for the angle θ.

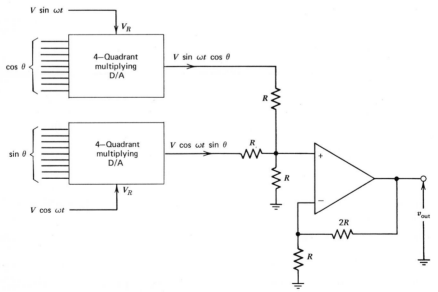

Figure 8.15 Digital phase shifter using two MDACs.

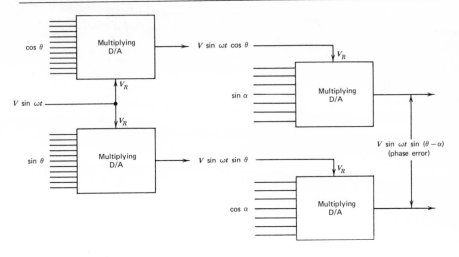

Figure 8.16 Phase-error detector. (Reprinted from *Analog–Digital Conversion Handbook*, with permission of Analog Devices, Inc., Norwood, Mass.)

Cascading four MDACs as in Figure 8.16 produces a phase-error detector. The output is proportional to the angular error, $\theta - \alpha$ (for small angular errors). Operated in this mode, the cascaded connection simulates a resolver-control transformer.

D/A converters also serve usefully as function generators. The availability of monolithic D/A converters and programmable read-only memories (PROM) allows the user the flexibility to produce virtually any output wave shape. Figure 8.17 displays typical connections for a function generator in which the PROMs have been programmed to produce a sinusoidal digital code for the D/A converter. Connected as shown, the circuit has 256 output steps per cycle, each with 256 possible levels. The frequency of the output is

$$f_o = \frac{f_{\text{clock}}}{256} \tag{8.6}$$

The circuit allows precise control of frequency, phase angle, starting transients, and transients when shifting frequency or phase angle. The output amplitude may be modulated or controlled by varying the reference voltage V_R. A true function generator may be produced by connecting a number of PROMs in parallel and selecting the appropriate PROM enable lines. Digital function generators such as that illustrated are useful for wave-form generation in test equipment, modulators-demodulators (modems), computer speech synthesis, and electronic music synthesis.

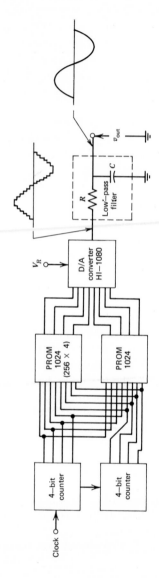

Figure 8.17 D/A function generator.

8.3 PRACTICAL LIMITATIONS OF D/A CONVERTERS

Most of the sources of error in D/A converters such as the finite "on" resistance and the voltage and current offsets of the ladder switches can be compensated for at room temperature in the design phase. Therefore, the drift in the device parameters with temperature changes is normally a major concern. Temperature coefficients associated with the reference-supply voltage and thermal matching among the ladder and feedback resistors become the limiting factors. Monolithic D/As are to be preferred over discrete or hybrid converters when significant temperature variations are to be encountered in the systems.

Whereas monolithic converters offer excellent thermal coupling among components, fabrication limitations and low yields restrict the number of usable input bits to about 10. Many D/A users would expect a 12-bit D/A to be accurate to 0.01% over the full temperature range quoted. This accuracy is equivalent to $\pm\frac{1}{2}$ LSB for the 12 bits. However, the converter is more likely to be accurate to $\pm\frac{1}{2}$ LSB for some small temperature range around 25°C, then accurate to $\pm\frac{1}{2}$ LSB for 11 bits for a while, then to 10 bits, and then to 9 bits. Thus it is possible for a 12-bit D/A to have only 10-bit accuracy to $\pm\frac{1}{2}$ LSB. As stated previously, resolution and accuracy are independent quantities.

8.4 PRINCIPLES OF A/D CONVERSION

The function of an analog-to-digital converter is to measure the amplitude of an analog input signal and produce a set of N digitally coded output signals to represent the input. The amplitude of the analog input signal generally is sampled at a frequency much higher than the maximum frequency component of interest in the input. The digital output from the A/D converter can be in serial or parallel format, or in a combination of both.

There are several ways of classifying the many different A/D converters. In *Analog-to-Digital/Digital-to-Analog Conversion Techniques*, D. F. Hoeschele (2) suggests three classifications, as follows:

One method is to separate them (A/D converters) as programmed or nonprogramed. In programed A/D converters the conversion process is performed in a given number of steps, with each step clocked to take a fixed time interval. The nonprogramed type of A/D converter may require a sequence of events to take place before the conversion is complete; however, this sequence is not in fixed time steps and depends only on the response time of the conversion circuitry.

Another way of classifying A/D converters would be to group them according to whether they are of the feedback or the open-loop type. In open-loop converters a direct comparison is made between the analog input voltage and a reference analog voltage or voltages. The result of the comparison is a generated digital word that is equivalent to the analog input. In closed-loop A/D converters, as the conversion process proceeds, an analog voltage generated internally as a function of a digital word in the A/D converter is fed back to one input of the comparator. This voltage is compared against the input analog voltage to be converted, and when the feedback voltage is equal to the input analog voltage the conversion is complete. The digital word in the A/D converter is then the digital equivalent of the analog input.

A third method of subdividing A/D converters ... is to separate them into two groups, depending on whether they are of the capacitor-charging type or the discrete voltage comparison type. The capacitor-charging A/D conversion process depends basically upon digitally encoding the time to charge a capacitor to some reference voltage value or to the value of the input analog voltage. Discrete voltage comparison A/D converters use a conversion process that depends basically upon the generation of discrete voltages whose levels are equivalent to digital words, and the comparison of these discrete voltage levels against the input analog voltage to determine the equivalent digital word. The generation of the discrete voltages could be simultaneous or sequential or a combination of the two.

Let us now examine three of the "discrete voltage comparison" A/D converters which are particularly well suited for monolithic integration. These are the counter-ramp, the up-down counter type, and the successive-approximation converters.

Counter-Ramp A/D Converter

A high-speed D/A converter can be the heart of several very useful A/D converter types. Figure 8.18 is a simplified diagram of a counter-ramp A/D converter that employs a D/A converter. A multiple-stage counter circuit is driven from a clock and the counter output drives the D/A converter, producing a staircase voltage ramp. When the D/A output voltage equals the analog input voltage, the comparator changes state; at that instant, the counter state represents the digital equivalent of the analog input.

A complete circuit for a counter-ramp converter and typical wave forms are illustrated in Figure 8.19. The digital circuits shown are 9300 types, but comparable circuits from other TTL families will work equally well if functional differences are taken into account.

Since the D/A converter normally has a negative output level, a positive-input signal is compared by resistive summation at one comparator input, using the summing resistor internal to the D/A which closely matches the D/A equivalent output resistance. The comparator is strobed with the clock to prevent any D/A switching spikes from

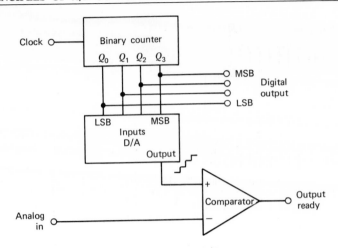

Figure 8.18 Counter-ramp A/D converter.

prematurely triggering the comparator. The strobing necessitates the use of the set–reset flip-flop formed by the cross-coupled gates, to make sure that the latch receives only one enable pulse per conversion cycle. Note that the data output from the latch is the complement of the digital value, because of the polarity conventions of the D/A.

To calibrate the A/D system for a 0 to +5 V input range, the input op amp is first zeroed in the conventional manner. Then +2.500 V is applied to the input, and the potentiometer between the reference supply and the D/A is adjusted so that the MSB just trips in at this level. The full-scale input will then be 1 LSB below +5.000 V, which is +4.980 V. Operation over a 0 to +10 V input range is easily accomplished by connecting the op amp output to the $2R$ sum terminal on the D/A. For a 0 to −5 V range, connect the op amp output to the negative input of the comparator through a 5 kΩ resistor. For −5 to +5 V bipolar operation, connect the ladder bus terminal on the D/A to the +5 V reference, and connect the reference through the potentiometer to the T_3 terminal of the D/A. Virtually any other input range is possible by changing the op amp gain or polarity, or adjusting the reference potentiometer. Zero shift may be accomplished by offsetting the ladder bus, or summing voltages at the op amp or comparator inputs.

System accuracy is affected primarily by the D/A accuracy, and to a lesser degree by offsets in the input op amp and the comparator. The circuit of Figure 8.19a proved to be accurate within $\pm\frac{1}{2}$ LSB at room temperature, and ±1 LSB from −55 to +125°C. This accuracy was maintained at clock rates up to 330 kHz. Clock rates up to 1 MHz could be used with about 1 additional LSB of inaccuracy.

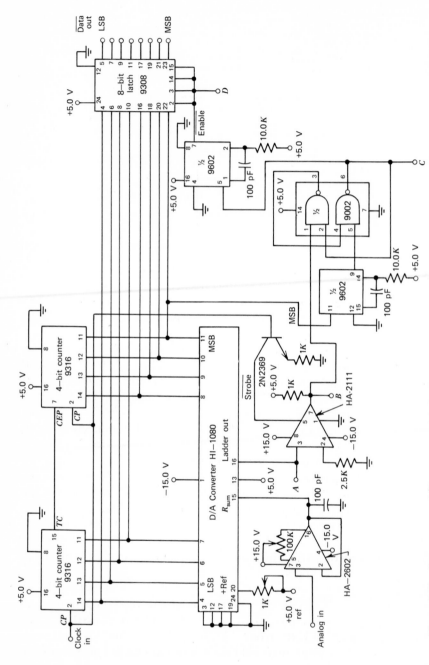

Figure 8.19 Counter-ramp A/D converter. (*a*) Complete circuit diagram. (*b*) Circuit wave forms with +2.50 V analog input.

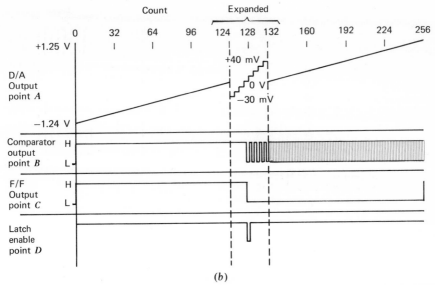

(b)

Figure 8.19 (cont.)

Using the illustrated circuit of Figure 8.19*a* as a starting point, many modifications to the digital circuitry are possible to suit the application. For example, if convert-on-command operation is desired, the circuitry beyond the comparator, including the latch, could be eliminated and the comparator output used to gate off the clock signal. The counters will hold their value until a command to convert again is issued by resetting the counters to zero. For continuous conversion, a reduction in average (but not maximum) conversion time can be made by resetting the counters immediately after data are entered in the latches.

Another possible improvement in conversion time can be achieved by running the clock at a variable rate—fast while the D/A output is far from the input level and slower when the comparator is about ready to trip. For example, one can use a voltage-controlled oscillator as the clock, since it will be controlled in frequency by an op amp with inputs wired across the comparator inputs. Another possibility would be to use a fixed 5 MHz clock and insert a divide-by-16 counter in series with the clock line when the D/A and input voltages are nearly equal. This arrangement could be controlled by a second comparator having the trip point offset from that of the main comparator.

Up-Down Counter for A/D conversion

Another A/D converter that uses a D/A is the up-down counter, or servo-type converter. This converter is most efficient in monitoring one

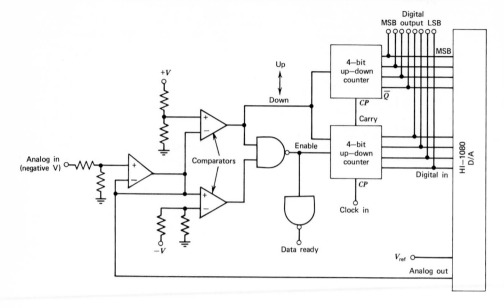

Figure 8.20   Up-down counter-type A/D converter.

analog signal continuously, rather than monitoring multiplexed analog signals. The converter works basically by balancing the input analog signal with the D/A output and adjusting the D/A by running a digital counter up or down, as required, to balance the signal. When the two analog signals balance, the counter state represents the digital equivalent of the input signal.

In the up-down counter converter illustrated in Figure 8.20, two analog signals are fed differentially into an op amp. For a positive-input signal, the D/A could be run in the positive-output mode, or in the negative-output mode by summing the two signals at the inverting amplifier input. The amplifier gain should be set at 2 or greater to allow less critical thresholds for the comparators. The two comparator thresholds are set up by voltage dividers to correspond to unbalances of approximately $\pm\frac{1}{2}$ LSB. When the analog signals are balanced within this range, the comparator outputs are both high, which stops the counters and gives a data ready signal to indicate that the digital outputs are correct. If the analog signals are unbalanced by more than $\pm\frac{1}{2}$ LSB, the counter is enabled and is driven in the up or down direction; depending on the polarity of the imbalance. The D/A converter in this application should be monotonic, meaning that the errors in two adjacent output steps should not cause a step in the wrong direction, for system oscillations may result.

If the D/A converter is operated in the negative output mode, the digital outputs will follow negative logic convention. If the analog input signal varies by less than 1 LSB per clock period, the converter will continuously track the signal.

The data ready signal could be useful in adaptive systems for most efficient data transfer, since that signal changes state only when there is a significant change in the analog input. When monitoring a slowly varying input, it would be necessary to read out the digital output only after a change has taken place. The data ready signal could trigger a flip-flop to flag this condition, and the flip-flop would be reset after read-out.

The main disadvantage of the up-down counter converter is the time required to initially acquire a signal—in an 8-bit system, as many as 256 clock periods might be needed. The input signal usually must be filtered to ensure that its rate of change does not exceed the tracking rate of the converter.

Successive-Approximation A/D Converter

Suppose you were trying to guess a number between 0 and 15 by asking the least number of questions answerable by yes or no. One of the most efficient ways might work as follows:

"Is it 8 or greater?"

"Yes"

"Is it 12 or greater?"

"No"

"Is it 10 or greater?"

"Yes"

"Is it 11?"

"No"

If you had jotted down a "1" for each "yes" and a "0" for each "no," you would have 1010, which of course is the binary notation for ten. Thus it is possible to find one number out of 16 with four questions. Likewise eight questions would be required to find a number between 0 and 255. Obviously this technique is usually much quicker than saying, "Is it zero?", "Is it one?," and so on, or guessing numbers at random.

Successive-approximation converters are the most popular A/D types

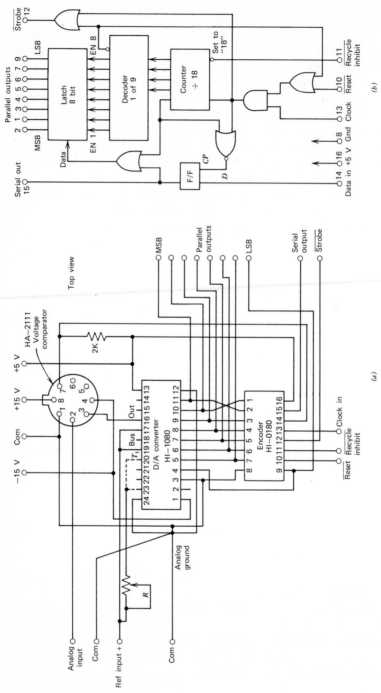

Figure 8.21 Parts of a successive-approximation A/D converter. (*a*) Complete circuit diagram showing three IC devices used. (*b*) Functional diagram of encoder IC.

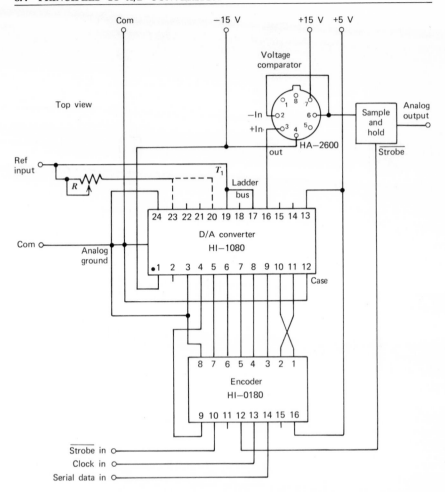

Figure 8.22 D/A converter for serial data input.

that utilize a D/A. Successive-approximation techniques are particularly useful for high-speed conversion, since only N clock cycles are needed for N-bit conversion. The counter-type converters previously discussed require up to 2^N cycles.

One disadvantage of implementing successive-approximation converters is that a sample-and-hold circuit is nearly always required in the signal path when only a few conversions per second are needed. The sample-and-hold circuit is an additional error source which is difficult to control over a wide temperature range. The counter-type converter does not require a sample-and-hold circuit, since its output is a parallel digital number taken

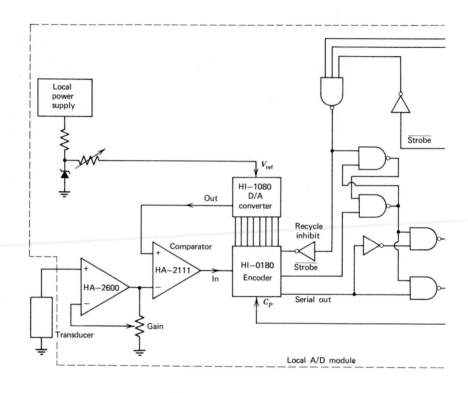

Figure 8.23 Data-acquisition system utilizing

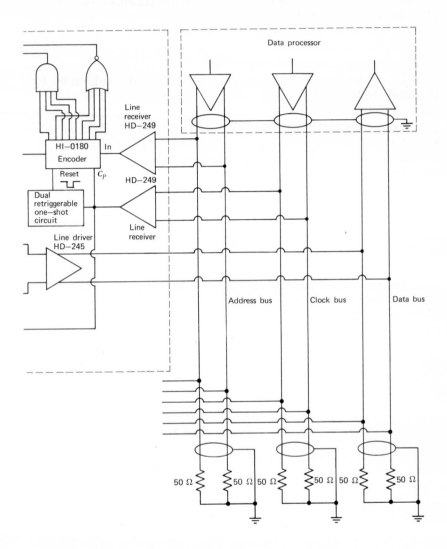

localized A/D module for each transducer.

at the instant that the D/A and input signals are equal, although filtering of the input signal may be desirable in some applications.

It can be seen that the 8-bit successive-approximation converter will give the correct output in eight clock cycles, whereas the up-down counter type could take as many as 255 cycles to acquire a signal. Once acquired, the up-down counter can indicate a change in a slowly varying signal within one clock cycle, but the successive-approximation type must step through another eight cycles. The choice really depends on the type of signals to be monitored.

The 8-bit successive-approximation A/D converter is easily implemented using three ICs (see Figure 8.21a). The input polarity and/or output logic polarity can be reversed by wiring modifications without additional components. Both serial and parallel outputs are available, plus a strobe output for synchronization. Conversions may be continuous, or controlled by external commands to the recycle inhibit pin. In addition to its use in data-acquisition systems, the converter is useful in digital filters, pulse-code modulation (PCM) telemetry, and digital voice communications. Using the D/A converter output as a subsystem output, the circuit becomes an analog sample-and-hold circuit with an arbitrarily long driftless time.

All the logic necessary to interface with the D/A converter has been integrated into the HI-0180 encoder, whose functional diagram appears in Figure 8.21b. This device contains all the digital circuitry necessary to interface with the D/A and replaces four conventional medium-scale integration IC devices. The other circuit that completes the successive-approximation converter is a precision voltage comparator. The low bias current of the HA-2111 allows it to be used without buffer amplifiers in many applications.

By changing the connection scheme between the three ICs, a D/A converter can be constructed. Figure 8.22 shows a D/A circuit useful at the receiving end of a serial digital transmission system. This circuit converts a *serial* digital input into an analog output.

Applications of Successive-Approximation A/D Converters   Most data-acquisition systems consist of transducers located at the points of measurement connected by analog signal lines to remotely located signal conditioners, an analog multiplexer, and a time-shared A/D converter. Since the signal lines from the points of measurement are particularly prone to noise pickup, these lines ought to be made digital. This has seldom been done, because many A/D converters would be required, and these have been expensive, bulky, and sensitive to the external environment. The single central A/D converter is usually a 10- to 16-bit device, grossly overspecified on the false assumption that this resolution will somehow overcome transducer errors (which are usually 1% or greater) and noise pickup on the analog lines. The central converter must also be high-speed to handle many multiplexed signals.

The alternate approach of locating an A/D converter next to each transducer is now quite practical. Using the three IC devices previously discussed, an 8-bit successive-approximation A/D converter can be constructed, capable of as many as 40,000 conversions/s with 1 LSB absolute accuracy over a −55 to +125°C ambient temperature range. Let us now discuss an interesting application of these interfacing devices.

In the data-acquisition system of Figure 8.23, a small card or module is located near each transducer, and a number of modules can be "party-lined" to three twisted, shielded cable pairs extending up to several thousand feet. The A/D module is normally "off-line" until addressed by the data processor. To start operation, the processor sends an 8-bit serial address down the address bus. This address is picked up in the module by a line receiver and loaded into a HI-0180 encoder, which is used as a serial-to-parallel converter. The gates connected to the parallel outputs of the encoder are wired uniquely for each module in the system; thus each module recognizes only its own address. If the correct address has been received, the line-driver circuit is enabled and the A/D converter is turned "on" for one conversion, the serial data being transmitted to the data bus. Periodically, all the modules are synchronized by interrupting the clock bus signal for a few cycles, which allows the dual retriggerable mono-stable circuit to fire, resetting the serial-to-parallel converter.

8.5   CONCLUSIONS

We have seen in this chapter how information can be converted from a digital format into an analog mode, and vice versa, using D/A and A/D converters. The terminology associated with these interfacing devices has been introduced, and the principles of operation have been examined. We have seen the utility of these converters through the selected applications of multiplying D/A converters, encoders, phase detectors, function generators, and data-acquisition systems.

BIBLIOGRAPHY

1. Dooley, D. J., "Applying a Monolithic 10-Bit D/A Converter," Application note for mono DAC-02, Precision Monolithics, Santa Clara, Calif.

2. Hoeschele, David, Jr., *Analog-to-Digital, Digital-to-Analog Conversion Techniques*, Wiley, New York, 1968.

3. Millman, J., and C. C. Halkias, *Integrated Electronics: Analog and Digital Circuits and Systems*, McGraw-Hill, New York, 1972.

4. Sheingold, Daniel H., *Analog-Digital Conversion Handbook*, Analog Devices Inc., Norwood, Mass., 1972.

CHAPTER 9

PHASE-LOCKED LOOPS

In electronics, we are concerned with the manipulation of information that is expressed as electrical quantities. This information can be expressed many ways—as a voltage, a current, or a time period between two events, or as modulation information on an ac carrier–amplitude, frequency, or phase angle.

Of these, the time-related expressions, time, frequency, and phase angle, have several advantages. Information can be transferred between remote points with less loss of accuracy due to noise. Extremely accurate standards are available—namely, crystal oscillators and atomic clocks.

One reason for the relative neglect by designers of frequency or phase modulation as information media is that hardware to manipulate these signals has been more complex than that for voltage and current signals. For voltage and current signals, inexpensive building blocks such as operational amplifiers and digital circuits have greatly simplified design of complex systems. Now there is a versatile building block for signals in the frequency domain—the monolithic phase-locked loop (PLL).

9.1 INTRODUCTION

Phase-locked loop concepts were first proposed more than 40 years ago by de Bellescize (4). De Bellescize investigated the subject of synchronous reception of radio signals when a receiver consisting of only a local oscillator, a mixer, and an audio amplifier was employed. The oscillator was adjusted to exactly the same frequency as the carrier frequency. With no frequency modulation of the carrier, an intermediate frequency of 0 Hz was created. However, when the carrier was frequency modulated, the output from the mixer represented the desired demodulated information carried by the signal.

The principal shortcoming of the receiver system as proposed by de Bellescize was the practical limitation of achieving perfect synchronization between the carrier frequency and the frequency of the local oscillator. Any frequency mismatch hopelessly garbled the information. To achieve the desired operation, the local oscillator and the transmitted signal had to be frequency- and phase-locked together with a degree of synchronization that was impractical at that time. Thus the more complicated but more practical technique of the superheterodyne receiver became the accepted approach for demodulation.

These early investigations by de Bellescize led to numerous systems employing phase-locked loops being designed to overcome the practical constraints that had plagued previous attempts. A few areas in which successful phase-locked loop operation has been achieved include the synchronization of horizontal and vertical scanning signals in television receivers, removal of the Doppler frequency shift in satellite tracking, stabilization of the frequency of klystron oscillators, and noise filtering in many communication systems. Virtually all these applications of PLLs employ a variety of sophisticated circuit techniques utilizing complex, inductively tuned filters to achieve the required frequency stability. Other potential application areas of PLLs exist where the use of this technique in conjunction with discrete components remained either too complex or too expensive to be justifiable until the advent of the integrated circuit PLL.

The recent development of the IC phase-locked loop has resulted in the removal of many of the practical limitations that prevented utilization of most of the simpler and more direct instrumentation techniques.

Synchronous detection without tuned circuits is now feasible and economically advantageous. Also, PLLs can now be employed economically in a variety of related applications, such as frequency synthesizers, tracking filters, motor speed monitoring and controls, modems, tone decoders, and frequency shift keying (FSK) receivers. In this chapter we describe the concept of PLL operation and examine several of these applications.

9.2 PRINCIPLES OF PHASE-LOCKED LOOPS

The term "phase-locked loop" is undoubtedly familiar, but unless one has had the opportunity to work with such a circuit, the concept may not be clear. The general operation is not difficult to understand, however. First, let us review what is meant by "loop"; later we can deal with the "phase lock."

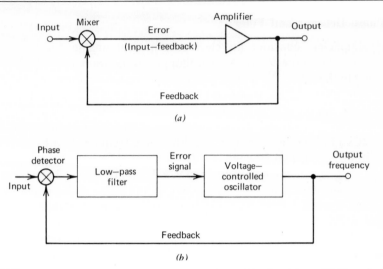

Figure 9.1   Comparing typical feedback loops. (a) General feedback loop. (b) Phase-locked loop.

Figure 9.1a is a general diagram that fits many control systems—electronic, mechanical, or even organic (i.e., the human body). The input signal is a function of the desired output. If the output is not presently at the desired point, an error signal is generated by the mixer, which is amplified and drives the output in a direction to minimize the error signal. The PLL works much the same way. Consider a simple diagram of a phase-locked loop (Figure 9.1b). If we consider the phase detector and low-pass filter to be a mixer, and the voltage-controlled oscillator to be a type of amplifier, the diagram is identical to the negative feedback loop in Figure 9.1a. In many feedback control systems, the input and output information signals are expressed as voltages. In the phase-locked loop, however, the input and output information signals are expressed as ac frequencies. The function of a PLL is to detect and track small differences in phase and frequency existing between an incoming signal and a secondary reference signal.

As in the general loop, the output is driven in the direction that will tend to minimize the error signal. In this case, the error signal is a frequency; thus the loop tends to drive the error frequency toward zero frequency. To accomplish this, the feedback frequency must be made equal to the input frequency. Once the two frequencies are made equal, the error signal is a function of phase difference between the two signals, and the phase difference therefore is also controlled. Before showing what we can do with the phase-locked loop, let us examine its parts.

Phase Detector and Filter

For simplicity, consider an electromechanical analog for the phase detector (see Figure 9.2a). As the chopper armature is moved from one contact to the other, the input signal appears at the armature alternately at 0 or 180° phase. If the input and feedback frequencies f_1 and f_2 are not equal, the circuit acts as a balanced modulator with sum and difference frequencies $(f_1 + f_2)$ and $(f_1 - f_2)$ appearing at the chopper armature. The low-pass filter will pass mainly the difference frequency with increasing amplitude as the two frequencies approach each other. If the input and feedback frequencies are exactly equal, a dc component will appear at the filter output, with amplitude dependent on the phase difference between the two signals. Figure 9.2b illustrates wave forms for the conditions under which $f_1 = f_2$ and phase difference of 0, 90, and 180°; Figure 9.2c shows the low-pass filter output as a function of phase difference between f_1 and f_2. In this electromechanical analog to the phase detector, we see that the output level is proportional to the input amplitude, as well as phase angle, and that zero output occurs at 90° phase angle. This is also true of most solid-state phase detectors, which actually multiply the two ac input signals together.

Voltage-Controlled Oscillator

The voltage-controlled oscillator (VCO) is usually an astable multi-vibrator with a dc input which can be used to vary the "free-running" frequency f_o' over a certain range. Figure 9.3a shows a simple VCO. Connecting the control input to $+V$ produces the conventional, fixed-frequency multivibrator. However, the application of a separate control input voltage alters the charging rate of the coupling capacitors. As the dc control input voltage becomes more positive, the capacitors charge more rapidly, and the output frequency increases. This frequency dependence is illustrated in Figure 9.3b.

9.3 PHASE-LOCKED LOOP OPERATION

Now we connect a signal generator to the PLL in Figure 9.4a and vary the input frequency, choosing the timing components in the VCO to produce the characteristic appearing in Figure 9.4b. The VCO can be constrained to run between 900 and 1100 Hz by limiting the voltage swing at point A. Since the detector output characteristic (Figure 9.2c) is bounded at both ends, the oscillator can only function between the limits determined by its input characteristics (Figure 9.4b) and the detector output characteristic.

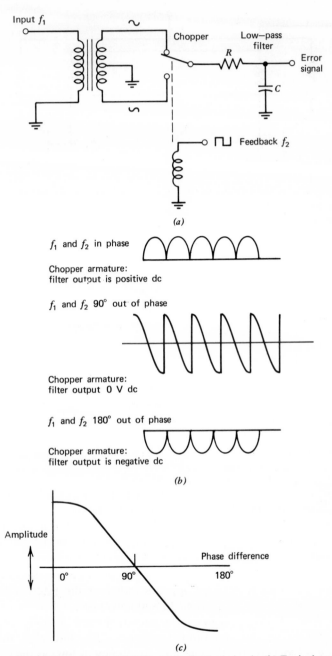

Input f_1

Chopper

Low–pass filter

R

Error signal

C

Feedback f_2

(a)

f_1 and f_2 in phase

Chopper armature:
filter output is positive dc

f_1 and f_2 90° out of phase

Chopper armature:
filter output 0 V dc

f_1 and f_2 180° out of phase

Chopper armature:
filter output is negative dc

(b)

Amplitude

Phase difference

0°   90°   180°

(c)

Figure 9.2 Phase detector analog. (a) Electromechanical circuit. (b) Typical wave shapes for different phase relationships. (c) Low-pass filter output as a function of phase difference.

333

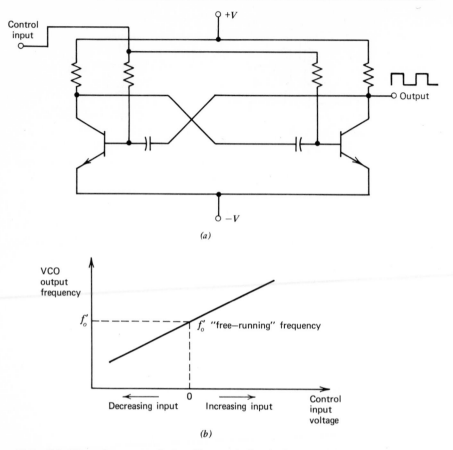

Figure 9.3 The voltage-controlled oscillator. (*a*) Circuit diagram. (*b*) Output frequency as function of input control voltage.

If the signal generator is disconnected, the VCO will run at $f_o' = 1000\,\text{Hz}$, since the filter output is zero volts. If we connect the signal generator with its frequency set to 700 Hz, the VCO will still run at f_o', since the difference frequency is high and very little signal gets through the low-pass filter. As we slowly increase the generator frequency, we will observe more and more "jitter" on the VCO output, as the difference frequency becomes lower and appears larger at the filter output, frequency modulating the VCO about f_o'. Suddenly—say, when the generator reaches about 920 Hz—the VCO frequency jumps abruptly and holds steady exactly at the generator frequency. We find that the VCO frequency is "locked" to the generator frequency as long as the generator remains between 900 and 1100 Hz. If the generator frequency goes outside these limits, the VCO will snap back to f_o'. As the generator approaches f_o' from either direction, we find that we have to go slightly closer to f_o' to initially achieve "lock" than the 900 to 1100 Hz limits required to hold the signal in "lock." These two sets of limits are known as *capture range* and *lock range*, respectively, and they can be independently adjusted by changing circuit constants in the PLL. Prior to lock, as the generator frequency approaches f_o' from either direction, the difference frequency from the phase detector becomes smaller and the amplitude passed though the filter becomes larger. This signal modulates the VCO frequency about f_o' until the VCO frequency sweeps past the input frequency; then "lock" takes place.

But why do the two signals "lock"? We can examine the stability of any closed-loop system by imagining what happens if the output takes a small random drift in either direction. Assume for the moment that the generator and VCO frequencies are identical. The low-pass filter output will be the dc level required to maintain the VCO at that frequency in accordance with Figure 9.4b. The phase difference between the VCO and the generator will not necessarily be 90°—it will be whatever phase angle is necessary to produce the proper VCO dc input through the phase

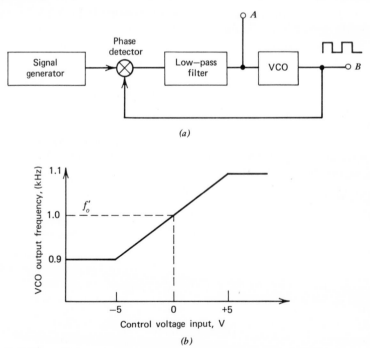

(a)

(b)

Figure 9.4 PLL example. (a) PLL block diagram. (b) VCO characteristic.

detector and filter to hold the oscillator frequency identical with the input frequency. Now suppose that the VCO frequency drifts upward from the generator frequency by a fraction of a cycle. The first indication of this event will be a change in the phase difference between the two signals—the VCO phase will become more leading. If the phase detector is connected such that a leading VCO phase will make its output less positive, the filter output will tend to pull the VCO back to its original frequency. Conversely, a tendency for the VCO to drift downward will also be corrected. For the same reasons, any change in the generator frequency will be tracked by the VCO frequency. Since the phase-detector output level as a function of phase angle is a cosine curve (Figure 9.2c), there will be two points in 360° where the detector output is at the proper level to drive the oscillator to the input frequency. But since only one of these points has positive stability, the loop will automatically settle on the stable side of the curve.

9.4   LINEAR MODEL FOR A PLL SYSTEM

Since the PLL is basically an electronic servo loop, many of the analytical techniques developed for control systems are applicable to phase-locked systems. Consider the basic PLL system presented in Figure 9.5a, where a current-controlled oscillator (CCO) has been substituted for the VCO. (These two types of oscillator operate in an identical manner except that the output frequency of the CCO is modulated by a dc input current rather than an input voltage. As we saw in Section 3.14, current control of oscillator frequency is preferred in integrated circuit design.)

Whenever a phase lock is established between $v_{in}(t)$ and $v_{out}(t)$, the linear model of Figure 9.5b can be used to predict the performance of the PLL system. Here ϕ_i and ϕ_o represent the phase angles associated with the input and CCO, respectively; $F(s)$ represents a generalized current transfer function of the low-pass filter in the frequency domain; K_d and K_o are the conversion gains of the phase detector and current-controlled oscillator, each having units as shown.

The representation of the phase detector as a summing network for combining $\phi_i(s)$ and $\phi_o(s)$ can be explained as follows. The phase detector is an analog multiplier that forms the product of an RF input signal $v_{in}(t)$ and the output signal, $v_{out}(t)$, from the current-controlled oscillator. (Basic circuits for and the operation of analog multipliers are covered in Section 3.13.) Assume that the two signals to be multiplied can be described by

$$v_{in}(t) = V_{in} \sin \omega_i t \qquad (9.1)$$

$$v_{out}(t) = V_{out} \sin (\omega_o t + \phi) \qquad (9.2)$$

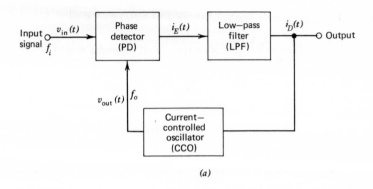

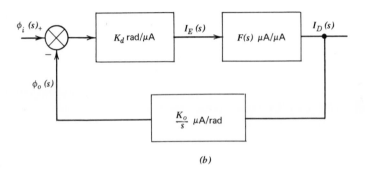

Figure 9.5 PLL system. (a) Block-diagram representation. (b) linear model.

where ω_i, ω_o, and ϕ are the frequency and phase characteristics of interest. The produce of these two signals is an output current signal given by

$$i_E(t) = K_1 V_{in} V_{out} (\sin \omega_i t)[\sin (\omega_o t + \phi)] \qquad (9.3)$$

where K_1 is an appropriate dimensional constant. Note that the amplitude of $i_E(t)$ is directly proportional to the amplitude of the input signal V_{in}. The two cases of $\omega_i \neq \omega_o$ and $\omega_i = \omega_o$ are of interest and are now considered separately.

Case I: $\omega_i \neq \omega_o$

When the two input frequencies to the phase detector are not synchronized, the loop is not locked. Furthermore, the phase angle ϕ is meaningless for this case, since it can be eliminated by appropriately choosing the time origin. The phase detector's output current signal is given by

$$i_E(t) = \frac{V_{in} V_{out} K_1}{2} [\cos (\omega_i - \omega_o)t - \cos (\omega_i + \omega_o)t] \tag{9.4}$$

When $i_E(t)$ is passed through the low-pass filter $F(s)$, the summed frequencies are attenuated, leaving

$$i_D(t) = V_{in} V_{out} K \cos (\omega_i - \omega_o)t \tag{9.5}$$

where K is a constant. Equation 9.5 represents the signal current supplied to the CCO to control its output frequency. The equation shows that this signal sets up a beat frequency between ω_i and ω_o, causing the CCO's frequency to deviate by $\pm\Delta\omega$ from ω_o in proportion to the signal amplitude ($V_{in} V_{out} K$) passing through the filter. If the amplitude of V_{in} is sufficiently large, and signal limiting or saturation does not occur, the CCO's output frequency will be shifted from the free-running frequency, ω_o', by some $\Delta\omega$ until lock is established where

$$\omega_i = \omega_o = \omega_o' \pm \Delta\omega \tag{9.6}$$

If lock cannot be established, then either (1) the input-signal amplitude is too small to drive the CCO to produce the necessary $\pm\Delta\omega$ deviation, or (2) ω_i is beyond the dynamic range of the CCO (i.e., $\omega_i \lessgtr \omega_o' \pm \Delta\omega$). Obvious remedies for these "no-lock conditions" are as follows:

1. Increase the input signal's amplitude. This can be done either internally or externally to the loop by employing additional amplification.
2. Increase the internal loop gain by adjusting upward (larger -3 dB frequency) the response of the low-pass filter.
3. Shift the free-running frequency ω_o', of the CCO closer to the expected ω_i frequency. Establishing frequency lock for the loop leads to the second case, namely, phase lock where $\omega_i = \omega_o$.

Case II: $\omega_i = \omega_o$

When ω_i and ω_o are frequency synchronized, the output signal from the phase detector for $\omega_i = \omega_o = \omega$, and for a phase shift of ϕ it is

$$i_E(t) = K_1 V_{in} V_{out} (\sin \omega t)(\sin \omega t + \phi)$$

$$= \frac{K_1 V_{in} V_{out}}{2} [\cos \phi - \cos (2\omega t + \phi)] \tag{9.7}$$

The low-pass filter removes the high-frequency, ac part of $i_E(t)$, leaving the dc component as a current signal for the CCO. Thus we have

$$i_D(t) = I_D = \frac{K_2 V_{in} V_{out}}{2} \cos \phi \qquad (9.8)$$

where K_2 is a dimensional constant.

Suppose ω_i and ω_o are perfectly synchronized to the free-running frequency ω_o'. For this case I_D will be zero, indicating that the ϕ must be 90°. Thus the error signal I_D is proportional to a phase difference between ω_i and ω_o centered about a reference phase angle of 90°. If ω_i now changes from ω_o', I_D will adjust and settle out to some nonzero value to correct ω_o; under these conditions, frequency lock is maintained with $\omega_i = \omega_o$. The ϕ will be shifted by some amount, $\Delta\phi$, from the reference phase angle of 90°. This concept can be simplified by redefining ϕ as

$$\phi = \phi_o \pm \Delta\phi \qquad (9.9)$$

where ϕ_o is the inherent 90° phase shift and $\Delta\phi$ represents departures from this reference value. Now the I_D becomes

$$I_D = V_{in} V_{out} K_2 \cos (\phi_o \pm \Delta\phi) = \pm V_{in} V_{out} K_2 \sin \Delta\phi \qquad (9.10)$$

Since the sine function is odd, a momentary change in $\Delta\phi$ contains information about which way to adjust the CCO frequency to correct and maintain the locked condition. The maximum range over which $\Delta\phi$ changes can be tracked is −90 to +90°. This corresponds to a ϕ range from 0 to 180°. In addition to being an error signal, I_D represents the demodulated output of an FM input applied as $v_{in}(t)$ assuming a linear CCO characteristic. Thus FM demodulation can be accomplished with the PLL without the inductively tuned circuits that are employed with conventional detectors. The useful frequency range over which the PLL can track an input signal is called the lock range or alternatively, the tracking range. The lock range is determined primarily by the maximum frequency swing possible in the CCO and the maximum output current from the phase detector for the input-signal level.

When the frequency of the input signal deviates from the reference frequency ω_o' of the CCO by more than the tracking range, the loop becomes unlocked and the linear model of Figure 9.5 no longer describes the system. When this happens, the CCO runs at ω_o', and the output current I_D becomes zero. Phase lock is reestablished when the input frequency approaches ω_o'. The frequency range over which the PLL can acquire an input signal is denoted as the capture range, which is set by the low-pass filter's characteristics. The capture range is limited to a value less than the lock range.

For normal, phase-lock operation, the free-running frequency, ω_o', is first established under zero input-signal conditions. A particular ω_o' is normally positioned in the center of the frequency band of interest, making it possible for both positive and negative deviations of the input-signal frequency about ω_o' to be detected by the PLL. Next the particular low-pass filter configuration required by design considerations is determined for the system. The significant factors are the necessary capture range for the $\pm\Delta\omega$ modulations expected, high attenuation of the $2\omega_i$ component from the phase detector, and the desired loop response to rapid changes in input frequency.

9.5 DETERMINING MODEL PARAMETERS FROM DEVICE CURVES

Now that we have developed a useful model for a PLL system as given in Figure 9.5b, let us attempt to determine the conversion gains K_d and K_o for the phase detector and the CCO, respectively. The value of these parameters will be important for loop stability in the applications we consider later. We evaluate K_d and K_o for a typical set of operating conditions to illustrate our approach.

Consider that a 10 mV rms input signal is to be applied to a PLL having a free-running frequency f_o' of 1 MHz. Figure 9.6 plots typical loop gain and oscillator frequency versus input-current performance. From Figure 9.6a, the slope of the loop-gain characteristic is approximately 1% $\Delta f/f_o'$ per phase angle about the 90° reference phase. As is obvious from the graph, the slope (and ultimately K_d) depends on the input-signal level. This is typical for signal amplitudes from a few microvolts to slightly above 10 mV rms, where nonlinear operation occurs because of signal-swing limitations. For operation with signal levels below 10 mV rms, simply scale the slope by a factor of $V_{in}/10$, where V_{in} is the expected input voltage in millivolts rms. The graph in Figure 9.6a was obtained under closed-loop conditions; a simple, passive, RC low-pass filter was included in the loop, whose transfer function was of the form

$$F(s) = \frac{1}{1 + \tau s} \qquad (9.11)$$

At dc we have

$$F(0) = 1 \qquad (9.12)$$

and the low-pass filter does not affect the overall loop gain. To employ some other type of low-pass filter whose transfer function is $F_1(s)$, the slope of the loop-gain characteristic would be multiplied by $F_1(0)$. Figure 9.6b shows the open-loop variation of the CCO frequency as a function of the controlling input current. The slope of this characteristic is approximately 1.2% $\Delta f/f_o'$ per microampere for input-current variations less than

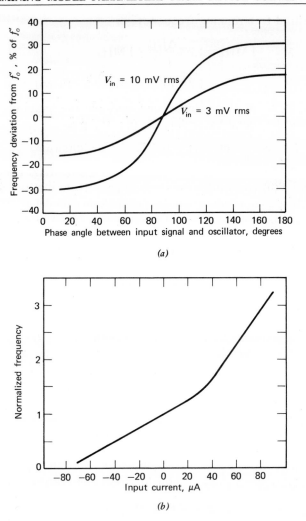

(a)

(b)

Figure 9.6 Characteristics of a PLL for determination of K_d and K_o conversion gains. (a) Loop-gain characteristic. (b) Oscillator frequency versus input-signal current.

30 μA. The conversion gain K_d is the quotient of these two slopes, or

$$K_d \cong \frac{\left(1\% \dfrac{\Delta f/f_o'}{\text{degree}}\right)}{1.2\% \dfrac{\Delta f/f_o'}{\mu A}} \cong 0.83 \ \mu A/\text{deg} = 48 \ \mu A/\text{rad} \qquad (9.13)$$

The K_o conversion gain of the CCO is the slope of Figure 9.6b multiplied by the desired free-running frequency, or

$$K_o \quad \text{at} \quad 1 \text{ MHz} = 1.2\% \frac{\Delta f / f_o'}{\mu \text{A}} \times 1 \text{ MHz}$$

$$= 1.2 \times 10^4 \text{ Hz}/\mu \text{A} = 7.5 \times 10^4 \text{ rad}/(\text{s})(\mu \text{A}) \quad (9.14)$$

The value of K_o at other frequencies between the limits of 1 kHz and 25 MHz may be found by scaling K_o according to

$$K_o \quad \text{at} \quad f_o' = (0.012)(2\pi f_o') \text{ rad}/(\text{s})(\mu \text{A}) \quad (9.15)$$

where f_o' is the free-running frequency of the CCO in MHz. The product of K_d, K_o, and $F(0)$ constitutes the dc loop gain. Note that the dimensions of $K_d K_o F(0)$ are reciprocal seconds. The dc loop gain defines the tracking range $\Delta\omega_T$, which is the range of frequencies centered about the free-running frequency over which the PLL can maintain lock. Note that additional dc gain can be designed into the PLL, allowing one or more amplifiers to be included in both the forward and feedback signal paths. The tracking range for the 1 MHz example is

$$\Delta\omega_T = \pm K_d K_o F(0) = \pm (48)(7.5 \times 10^4)(1) = \pm 3.6 \times 10^6 \text{ rad/s} \quad (9.16)$$

This $\Delta\omega_T$ corresponds to a frequency deviation of ± 570 kHz, indicating that the PLL can track input frequencies from approximately 430 kHz to 1.5 MHz. The $\Delta\omega_T$ expression is valid as long as nonlinear operation due to saturation or limiting does not occur within the loop, thereby producing a smaller value of $\Delta\omega_T$.

The frequency characteristics of the low-pass filter determine the sensitivity of the PLL by setting the capture range, $\Delta\omega_C$, which is the band of frequencies centered at f_o' at which the PLL can move from the unlocked state to establish or acquire lock with an input signal. The capture range is always less than the lock range and can be estimated by

$$\Delta\omega_C \approx K_o K_d \, |F(j \, \Delta\omega_C)| \quad (9.17)$$

where $|F(j \, \Delta\omega_C)|$ represents the magnitude of the low-pass filter at a radian frequency, ω, equal to the capture range $\Delta\omega_C$.

For a simple, passive low-pass filter of the form

$$F(s) = \frac{1}{1 + \tau s} \quad (9.18)$$

where the filter time constant τ is chosen such that

$$\tau \gg \frac{1}{2 \, \Delta\omega_T} \quad (9.19)$$

The capture range can be approximated as

$$\Delta\omega_C \approx \pm\sqrt{\frac{\Delta\omega_T}{\tau}} \tag{9.20}$$

As a practical example, suppose our low-pass filter is chosen such that

$$\tau = RC = (10\ \text{k}\Omega)(200\ \text{pF}) = 2\ \mu\text{s} \tag{9.21}$$

At $f'_o = 1$ MHz, we have

$$\Delta\omega_C \cong 1.3 \times 10^6\ \text{rad/s} \tag{9.22}$$

giving a capture frequency range of approximately 200 kHz. Thus for $f'_o = 1$ MHz with this filter, the frequency of the input signal must be in the range between approximately 800 kHz and 1.2 MHz to establish phase lock. However, once lock is established, the PLL system will track all input frequencies between 430 kHz and 1.5 MHz. The capture range may be adjusted between the limits of 0 to $\Delta\omega_T$ by varying the filter time constant τ. An active low-pass filter like that in Figure 9.7 may be employed in the PLL system. The response of this filter is identical to that of the passive filter except for the gain term R_1/R_2. By varying the forward current gain of the filter, the loop gain can be adjusted to control both the tracking range $\Delta\omega_T$ and the capture range $\Delta\omega_C$.

Certain PLLs are equipped with additional circuitry to enable independent adjustment of $\Delta\omega_C$ and $\Delta\omega_T$. Internal bandwidth control of $\Delta\omega_T$ is provided by a clipping network and unity-gain amplifier (see Figure 9.8). Figure 9.8 shows a current-clipping stage together with its current-transfer function. The current I_D represents the dc component of the output signal from the phase detector; I_{CCO} is the input current to the CCO which adjusts its frequency about f'_o; R_{CCO} is the input resistance of the CCO. The clipper diodes short when

$$V_D - V_{CCO} \cong \pm 750\ \text{mV} \tag{9.23}$$

limiting I_{CCO} to maximum and minimum values of $\pm 750/R\ \mu\text{A}$ (R in kilohms). By this clipping action, the bandwidth of the CCO is limited to

$$\Delta\omega_{CCO} = \pm\frac{750}{R}\ K_o\ \text{rad/s} \tag{9.24}$$

Evaluating $\Delta\omega_{CCO}$ for the previous 1 MHz example with $R = 10$ kΩ gives

$$\Delta\omega_{CCO} = \pm 5.6 \times 10^6\ \text{rad/s} \tag{9.25q}$$

or a possible CCO frequency change of ± 900 kHz. Since $\Delta\omega_{CCO} > \Delta\omega_T$ for this $R = 10$ kΩ, the tracking range is controlled by the dc loop gain and not this clipping network. If R were adjusted to 100 kΩ, the tracking-range

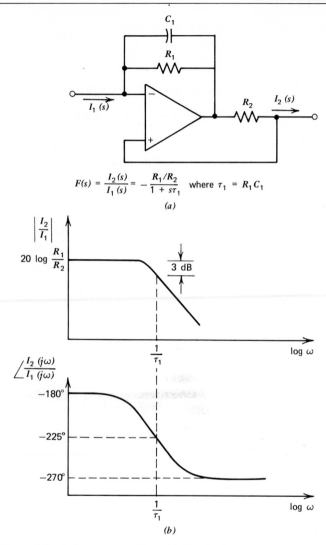

Figure 9.7   Active low-pass filter. (a) Circuit diagram. (b) Bode plot.

bandwidth would be set by the clipping network to 90 kHz. Figure 9.9 shows the typical variation of the tracking range for three different R settings. The clipping network also controls the maximum detectable phase shift in the loop. The phase detector's output current is

$$I_D = K_d(\phi_i - \phi_o) \tag{9.26}$$

where ϕ_i and ϕ_o are the phase angles of the input and CCO signals,

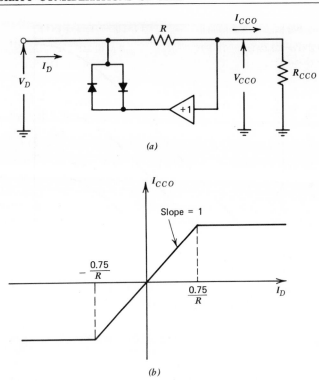

Figure 9.8 Current-clipping network for independent adjustment of $\Delta\omega_r$. (a) Circuit diagram. (b) Transfer characteristic.

respectively. Current limiting in the clipper of I_{CCO} gives a maximum phase shift of

$$(\phi_i - \phi_o)_{max} = \pm\frac{750}{RK_d} \text{ (rads)} \qquad (9.27)$$

The free-running frequency of the current-controlled oscillator is determined by the value of the timing capacitance C_T. Figure 9.10 gives typical C_T values for setting f'_o over its specified range of 1 kHz to 25 MHz.

9.6 STABILITY CONSIDERATIONS OF PLLs

Unexpected oscillations are always a possibility whenever feedback is employed to close a loop containing active elements. Both the stability and the operation of the PLL depend principally on the characteristics of the low-pass filter. A stability analysis using root-locus techniques will be given for several typical low-pass filters. The selectivity (i.e., tracking and

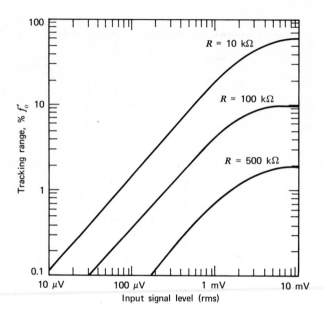

Figure 9.9   Tracking range versus input-signal level.

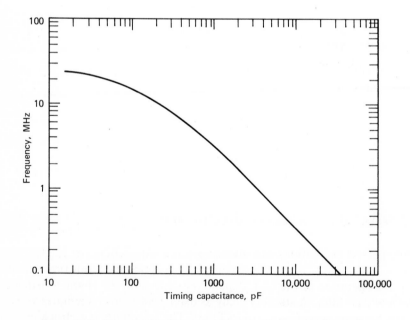

Figure 9.10   Free-running frequency f_o' versus timing capacitance C_T.

cature ranges), as well as the time response to changes in input frequency or phase angle, depend on the frequency characteristics of this low-pass filter.

Any linear low-pass filter has a generalized response given by

$$F(s) = \frac{\sum_{i=0}^{m} a_i s^i}{\sum_{j=0}^{n} b_j s^j} \tag{9.28}$$

where $m \leq n$. In typical filter applications, one particular pole will dominate in the $F(s)$ expression—the lowest of the high-frequency poles. As the location of this pole is moved closer to the zero-frequency origin, the frequency selectivity of the PLL to the $\omega_i - \omega_o$ difference increases. This increase in frequency selectivity naturally reduces both the tracking and capture ranges.

When the PLL is tracking an input signal, the loop gain is given by

$$G(s)H(s) = \frac{K_d K_o F(s)}{s} \tag{9.29}$$

The simplest low-pass current filter is the RC network in Figure 9.11a, which has a current transfer function given by

$$F(s) = \frac{I_2(s)}{I_1(s)} = \frac{1}{1 + \tau_f s} \tag{9.30}$$

where $\tau_f = RC$.

With this filter, the loop gain is

$$G(s)H(s) = \frac{K_d K_o}{s(1 + \tau_f s)} = \frac{K}{s(1 + \tau_f s)} \tag{9.31}$$

where

$$K = K_d K_o \tag{9.32}$$

The root locus for the PLL with the simple RC filter appears in Figure 9.11c. This locus shows the behavior of the PLL when the loop is closed for all positive values of gain K. The locus begins on the poles, corresponding to $K = 0$, and proceeds along the real axis until the critical value of gain is reached. For $K = K_{crit}$, the closed-loop response has the form

$$\frac{\phi_o(s)}{\phi_i(s)} = \frac{\dfrac{K_{crit}}{\tau_f}}{s^2 + \dfrac{s}{\tau_f} + \dfrac{K_{crit}}{\tau_f}} = \frac{\dfrac{K_{crit}}{\tau_f}}{\left(s + \dfrac{1}{\tau_{crit}}\right)^2} \tag{9.33}$$

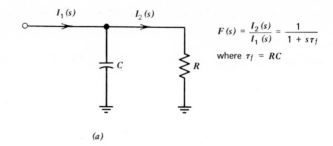

$$F(s) = \frac{I_2(s)}{I_1(s)} = \frac{1}{1 + s\tau_f}$$

where $\tau_f = RC$

(a)

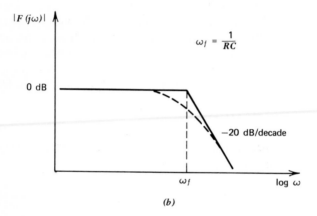

$$\omega_f = \frac{1}{RC}$$

(b)

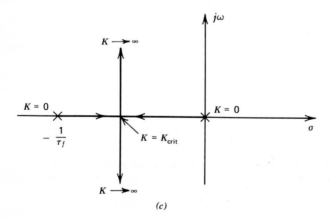

(c)

Figure 9.11 A passive, low-pass current filter. (*a*) Circuit configuration. (*b*) Frequency characteristics. (*c*) Root locus of PLL system.

348

This equation shows that the system is critically damped for this gain. When $K > K_{crit}$, the system will exhibit a damped sinusoidal response to an input phase-angle step function. Since the root locus always lies in the left half of the complex plane, the system is unconditionally stable for all K values.

In most phase-locked loop systems, the conversion gains K_o and K_d are functions of parameters internal to the integrated circuit, therefore not easily adjustable to achieve the desired system response. To overcome this shortcoming, an amplifier whose gain, A, is easily adjusted, can be inserted in the loop. When a typical amplifier with a single high-frequency pole is employed, the loop gain becomes

$$G(s)H(s) = \frac{K_o K_d A}{s(1 + s\tau_f)(1 + s\tau_a)} = \frac{K_1}{s(1 + s\tau_f)(1 + s\tau_a)} \qquad (9.34)$$

where A and τ_a represent the amplifier's characteristics. The root locus describing this loop gain is presented in Figure 9.12. In constructing this plot, the pole due to the amplifier was arbitrarily made larger than the filter's pole. However, the general shape of the root locus is independent of the relative positions of these two poles. This locus illustrates that the system will become unstable for all values of gain K_1 greater than K_{crit}. However, stable operation can be achieved by limiting the maximum gain of the amplifier (as well as $K_o K_d$) to a value less than that which produces K_{crit}.

Stable PLL operation with adjustable gain can be achieved by utilizing a filter having both a pole and a zero, like the filter in Figure 9.13. When

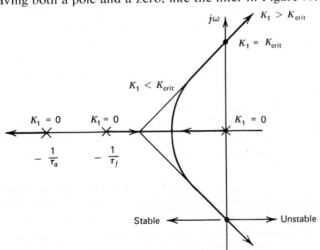

Figure 9.12 Root locus showing effect of amplifier pole.

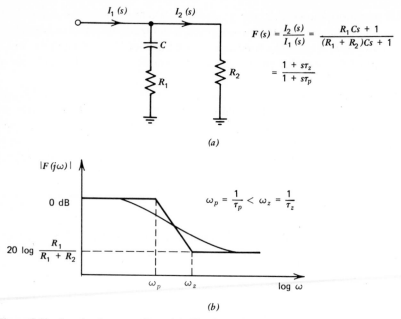

(a)

(b)

Figure 9.13 Lag–lead current filter. (*a*) Circuit configuration. (*b*) Frequency response.

this filter and the previous amplifier are employed, the loop gain becomes

$$G(s)H(s) = \frac{K_oK_dA(1+s\tau_z)}{s(1+s\tau_p)(1+s\tau_a)} \qquad (9.35)$$

A number of root loci are possible for this loop-gain expression, depending on the relative locations of τ_z, τ_p, and τ_a. Figure 9.14 illustrates all possibilities and indicates that the loop is now stable for all values of gain. Introduction of the zero has the effect of pulling the root locus to the left, away from the right-half plane where instability would result. The case in Figure 9.14*e*, showing pole-zero cancellation between the amplifier and the low-pass filter, is of particular interest because it results in the same root locus given in Figure 9.11*c*. Many other filter configurations for PLL operation could be examined if space permitted. However, the utilization of active filters with PLL offers many especially interesting possibilities; one example follows.

Figure 9.15*a* indicates the voltage- and current-transfer responses of a typical, generalized circuit configuration for an active filter. The transfer responses were obtained assuming an ideal operational amplifier ($A_v \to \infty$, $Z_{in} \to \infty \Omega$, $Z_{out} \to 0 \ \Omega$). The current-transfer function is particularly significant because it indicates that the output current is not a function of the load impedance Z_L. Thus in addition to being an active filter, this network can be made to supply a constant current to a varying load impedance. As a special example of the active filter of Figure 9.15, assume that impedances are chosen according to

$$Z_1 = 0 \qquad Z_2 = R_2 + \frac{1}{sC_2}$$

$$Z_3 = R_3 \qquad Z_L = R_L \tag{9.36}$$

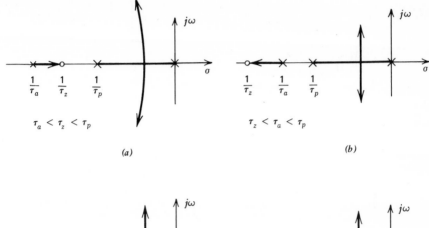

(a)

(b)

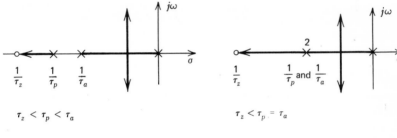

(c)

(d)

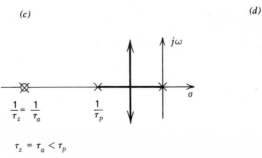

(e)

Figure 9.14 Root loci possibilities for lag–lead filter. (a) Zero between poles. (b) Zero to left of both poles with $\tau_a < \tau_p$. (c) Zero to left of both poles with $\tau_p < \tau_a$. (d) Equal poles. (e) Pole–zero cancellation.

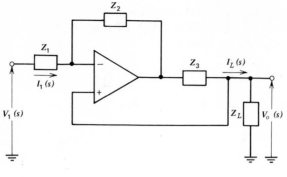

$$\frac{V_o(s)}{V_1(s)} = \frac{Z_L Z_2}{Z_L Z_2 - Z_1 Z_3}$$ Forward voltage transfer function

$$\frac{I_L(s)}{I_1(s)} = -\frac{Z_2}{Z_3}$$ Forward current transfer function

(a)

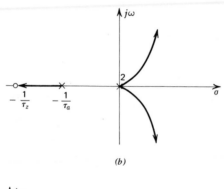

(b)

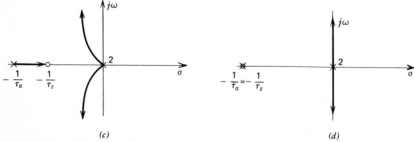

(c)   (d)

Figure 9.15   Active, low-pass current filter. (a) Circuit configuration and root locus of PLL with active filter, where (b) Zero is to left of amplifier's pole. (c) Pole is to left of active filter zero, and (d) Pole-zero cancellation.

Chosen thus, the response of the active filter is given by

$$F(s) = \frac{I_L(s)}{I_1(s)} = -\frac{1}{R_3 C_2}\left(\frac{1+s\tau_z}{s}\right) = K_f\left(\frac{1+s\tau_z}{s}\right) \qquad (9.37)$$

where $\tau_z = R_2 C_2$. For very small τ_z, the low-pass filter is essentially an integrator. When this filter is employed in the PLL, the loop again becomes

$$G(s)H(s) = \frac{K_o K_d K_f A (1+s\tau_z)}{s^2(1+s\tau_a)} = \frac{K_2(1+s\tau_z)}{s^2(1+s\tau_a)} \qquad (9.38)$$

Three root loci are possible depending on the relative locations of τ_z to τ_a, (see Figures 9.15b to 9.15d). Clearly, stability is possible only for Figure 9.15c, where the zero of the filter lies between the poles. Figure 9.15d, where the locus coincides with the imaginary axis, is considered "marginally stable." However, it is not useful practically.

Selecting various combinations of Z_2 and Z_3 impedances in the active filter of Figure 9.15a will allow us to synthesize a number of specialized transfer functions for particular applications. Table 1 of Figure 9.16 shows the Z_2 and Z_3 circuit configurations necessary for realizing special forms of $F(s)$. Each of the first four filters in the table is a single-pole filter and produces a second-order system response when included in the PLL. The $\Delta\omega_C$ and $\Delta\omega_T$ ranges can all easily be calculated for the first four cases using the expressions given in Table 2. The last two filters in Table 1 produce third-order systems that may be necessary for the PLL to track certain ramp- and acceleration-type inputs. The serious reader will be interested in the detailed treatments provided by Gardner (5) and Viterbi (16, 17), of the dynamic response, noise properties, loop stability, steady-state error, and other properties of the PLL when the various filters of Table 1 are used. A portion of Gardiner's work is particularly useful here and is reproduced for convenience in Table 3. This table contains a group of formulas for determining the transient phase error of a second-order loop for inputs of

1. A step of phase, $\Delta\theta$ rad

$$\left[\theta_i(s) = \frac{\Delta\theta}{s}\right] \qquad (9.39)$$

2. A step of frequency (phase ramp), $\Delta\omega$ rad/s

$$\left[\theta_i(s) = \frac{\Delta\theta}{s^2}\right] \qquad (9.40)$$

(b)

CASE	CAPTURE RANGE Δw_C	TRACKING RANGE Δw_T *
I	$\pm\, 1/\tau_2$	$\pm\, K_d K_0 R_2 / R_3$
II	$\pm\, \dfrac{1 + (R_{21}/R_3) K_0 K_d \tau_{20}}{\tau_{20} + \tau_{21}}$	$\pm\, K_0 K_d R_{21} / R_3$
III	0	∞
IV	$K_d K_0 \tau_2 / \tau_{23}$	∞

* The tracking range will be determined by the smaller value as determined by these equations compared with the value set by the clipping network.

CASE	Z_2	Z_3	$F(s)$	COMMENTS
I	$R_2 \parallel C_2$	R_3	$-\dfrac{R_2/R_3}{1+s\tau_2}$	$\tau_2 = R_2 C_2$
II	$R_{20},\ C_2,\ R_{21}$	R_3	$-\dfrac{R_{21}}{R_3}\dfrac{(1+s\tau_{20})}{1+(\tau_{20}+\tau_{21})s}$	$\tau_{20}=R_{20}C_2$ $\tau_{21}=R_{21}C_2$
III	C_2 †	R_3	$\dfrac{1}{\tau_{23}\,s}$	$\tau_{23}=R_3 C_2$ IDEAL INTERGRATOR
IV	$R_2,\ C_{21},\ C_2$ †	R_3	$-\dfrac{1+\tau_2 s}{\tau_{23}\,s}$	$\tau_2 = R_2 C_2$ $\tau_{23}=R_3 C_2$
V	$R_2,\ C_2,\ C_{21}$ †	R_3	$\dfrac{1+(\tau_2+\tau_{21})s}{\tau_3 s(1+s\tau_2)}$	$\tau_2 = R_2 C_2$ $\tau_{21}=R_2 C_{21}$ $\tau_3 = R_3 C_{21}$
VI	$R_2,\ C_2$ †	$R_3,\ L_3$	$-\dfrac{(1+s\tau_2)(1+s\tau_3)}{s^2 C_2 L_3}$	$\tau_2 = R_2 C_2$ $\tau_3 = L_3/R_3$

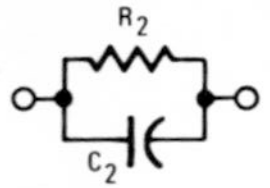

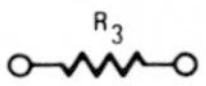

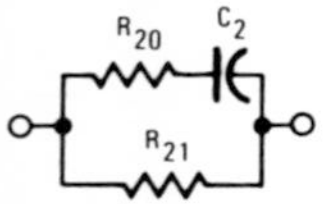

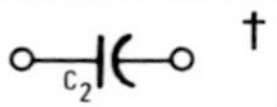

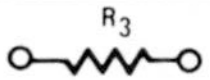

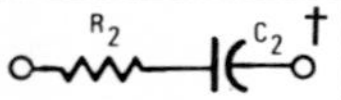

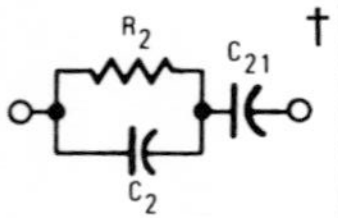

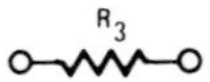

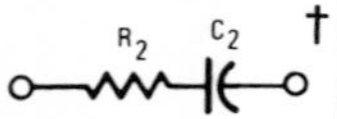

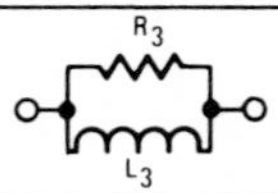

(a)

† Practical considerations will require shunting Z_2 with a large resistance in order to insure the stability of the op amp.

Figure 9.16 Tables for determination of PLL response. (a) Table 1, synthesis of active, low-pass current filters. (b) Table 2, formulas for determining PLL ranges. (c) Table 3, transient phase error of second-order loop.

354

	PHASE STEP ($\Delta\theta$ RADIANS)	FREQUENCY STEP ($\Delta\omega$ RAD/SEC)	FREQUENCY RAMP ($\Delta\dot\omega$ RAD/SEC^2)
$\zeta<1$	$\Delta\theta\left(\cos\sqrt{1-\zeta^2}\,\omega_n t - \dfrac{\zeta}{\sqrt{1-\zeta^2}}\sin\sqrt{1-\zeta^2}\,\omega_n t\right)e^{-\zeta\omega_n t}$	$\dfrac{\Delta\omega}{\omega_n}\left(\dfrac{1}{\sqrt{1-\zeta^2}}\sin\sqrt{1-\zeta^2}\,\omega_n t\right)e^{-\zeta\omega_n t}$	$\dfrac{\Delta\dot\omega t}{K_v}+\dfrac{\Delta\dot\omega}{\omega_n^2}-\dfrac{\Delta\dot\omega}{\omega_n^2}\left(\cos\sqrt{1-\zeta^2}\,\omega_n t +\dfrac{\zeta}{\sqrt{1-\zeta^2}}\sin\sqrt{1-\zeta^2}\,\omega_n t\right)e^{-\zeta\omega_n t}$
$\zeta=1$	$\Delta\theta(1-\omega_n t)e^{-\omega_n t}$	$\dfrac{\Delta\omega}{\omega_n}(\omega_n t)e^{-\omega_n t}$	$\dfrac{\Delta\dot\omega t}{K_v}+\dfrac{\Delta\dot\omega}{\omega_n^2}-\dfrac{\Delta\dot\omega}{\omega_n^2}(1+\omega_n t)e^{-\omega_n t}$
$\zeta>1$	$\Delta\theta\left(\cosh\sqrt{\zeta^2-1}\,\omega_n t - \dfrac{\zeta}{\sqrt{\zeta^2-1}}\sinh\sqrt{\zeta^2-1}\,\omega_n t\right)e^{-\zeta\omega_n t}$	$\dfrac{\Delta\omega}{\omega_n}\left(\dfrac{1}{\sqrt{\zeta^2-1}}\,\text{SINH}\sqrt{\zeta^2-1}\,\omega_n t\right)e^{-\zeta\omega_n t}$	$\dfrac{\Delta\dot\omega t}{K_v}+\dfrac{\Delta\dot\omega}{\omega_n^2}-\dfrac{\Delta\dot\omega}{\omega_n^2}\left(\cosh\sqrt{\zeta^2-1}\,\omega_n t +\dfrac{\zeta}{\sqrt{\zeta^2-1}}\sinh\sqrt{\zeta^2-1}\,\omega_n t\right)e^{-\zeta\omega_n t}$
	STEADY-STATE ERROR = 0	STEADY-STATE ERROR = $\dfrac{\Delta\omega}{K_v}$ (NOT INCLUDED ABOVE)	STEADY-STATE ERROR = $\dfrac{\Delta\dot\omega}{K_v}+\dfrac{\Delta\dot\omega}{\omega_n^2}$ (INCLUDED ABOVE)

(c)

Figure 9.16 (cont.)

3. A step of acceleration (frequency ramp), $\Delta\omega$ rad/s^2

$$\left[\theta_i(s)=\frac{\Delta\theta}{s^3}\right] \tag{9.41}$$

The phase error θ_e is the output from the phase detector and is expressed by

$$\theta_e(s)=\theta_i(s)-\theta_o(s)=\frac{s\theta_i(s)}{s+K_oK_dF(s)} \tag{9.42}$$

As Table 3 shows, the transient-phase error is a strong function of the undamped natural frequency ω_n and the damping ratio δ. These factors of the second-order equation are, in turn, set by the $F(s)$ of the low-pass filter.

In certain PLL applications, it may be desirable to use one type of low-pass filter for capturing the input signal and another filter for signal tracking. The four-channel programmable amplifier (HA-2400 PRAM) discussed in Chapter 6 functions ideally as a switchable active filter. Any one of the four operational amplifiers contained in this device can be activated and inserted in the loop by applying an external enable signal. Since different feedback circuitry can be used in conjunction with each op amp, as many as four different filters can be switched into and out of the loop to shape the PLL response to specifications.

9.7 STEADY-STATE RESPONSE OF THE PLL

The steady-state response of a phase-locked loop system to stimulation afforded by various types of input signal is most important to the system's stability. Although use of a certain low-pass filter frequently guarantees stability, it may produce a system with unsatisfactory steady-state characteristics for the types of input signal expected. The development that follows will be useful for determining the type of low-pass filter needed to enable the system to respond to step-function, ramp, and acceleration inputs.

With reference again to the linear model of the PLL system of Figure 9.5b, the current signal $I_E(s)$ represents the phase error that exists between the incoming reference signal $\phi_i(s)$ and the feedback signal $\phi_o(s)$. The final-value theorem associated with Laplace transformations, namely

$$\lim_{t\to\infty}[i_E(t)]=\lim_{s\to 0}[sI_E(s)] \tag{9.43}$$

will be used to find the steady-state error without transforming back to the time domain. The phase-error signal is given by

$$I_E(s) = \frac{K_d s}{s + K_d K_o F(s)} \phi_i(s) \tag{9.44}$$

For a unit step-function input

$$\phi_i(s) = \frac{1}{s} \tag{9.45}$$

For a unit ramp input

$$\phi_i(s) = \frac{1}{s^2} \tag{9.46}$$

For a unit acceleration input

$$\phi_i(s) = \frac{1}{s^3} \tag{9.47}$$

The steady-state responses of the error signal for these three inputs are as follows:

STEP: $$\lim_{t \to \infty} [i_E(t)] = \lim_{s \to 0} \left[\frac{K_d s}{s + K_d K_o F(s)} \right] \tag{9.48}$$

RAMP: $$\lim_{t \to \infty} [i_E(t)] = \lim_{s \to 0} \left[\frac{K_d}{s + K_d K_o F(s)} \right] \tag{9.49}$$

ACCELERATION: $$\lim_{t \to \infty} [i_E(t)] = \lim_{s \to 0} \left\{ \frac{K_d}{s[s + K_d K_o F(s)]} \right\} \tag{9.50}$$

As long as the filter is truly low-pass; that is, as long as it has a general response given by

$$F(s) = \frac{\sum_{i=0}^{m} a_i s^i}{\sum_{j=0}^{n} b_j s^j} \tag{9.51}$$

where $m \le n$, the following simplifications are possible:

STEP: $$\lim_{t \to \infty} [i_E(t)] = 0 \qquad \text{for any } F(s) \tag{9.52}$$

RAMP: $$\lim_{t \to \infty} [i_E(t)] = \frac{1}{K_o} \left\{ \lim_{s \to 0} \left[\frac{1}{F(s)} \right] \right\} \tag{9.53}$$

ACCELERATION: $$\lim_{t \to \infty} [i_E(t)] = \frac{1}{K_o} \left\{ \lim_{s \to 0} \left[\frac{1}{sF(s)} \right] \right\} \tag{9.54}$$

INPUT	TYPE 0	TYPE 1	TYPE 2
STEP	ZERO	ZERO	ZERO
RAMP	CONSTANT	ZERO	ZERO
ACCELERATION	CONTINUALLY INCREASING	CONSTANT	ZERO

Figure 9.17 Steady-state errors of various filter types for three different inputs.

Thus the PLL will track a step input with zero steady-state error regardless of the low-pass filter. To achieve zero steady-state error for an acceleration input, the low-pass filter must have at least two poles positioned at the origin (type 2 filter). Figure 9.17 shows the expected steady-state error for the different inputs when type 1, type 2, and type 3 low-pass filters are employed.

9.8 APPLICATIONS OF PLLs

Monolithic PLLs are available which work over an extremely broad band of frequencies from approximately 0.01 Hz to 25 MHz. In this section we examine several applications for two specific PLLs whose functional diagrams appear in Figure 9.18. The free-running frequency of the PLL in Figure 9.18a can be set anywhere between 0.01 Hz to 3 MHz; Figure 9.18b illustrates a PLL for which $5\ \text{MHz} \leqslant f'_o \leqslant 25\ \text{MHz}$. Figures 9.19 and 9.20 are complete schematic diagrams for both PLLs. Since the operation of these PLLs is quite similar, we will briefly explain the circuit operation of the HA-2800. Identification of the circuit functions will provide a better background for our discussion of specific applications.

As explained previously, the phase detector is basically an analog multiplier. The circuit multiplies by using the transconductance principle. The input signal applied at pins 7 and 10 produces ac currents i_1 and i_2 which are directly proportional to the applied input signal, or

$$i_1 = K_1 v_x = -i_2 \tag{9.55}$$

where K_1 is a constant and v_x represents the ac input voltage. Changing i_1 in accordance with v_x changes the effective input impedance of the difference amplifier composed of Q_{10}, Q_{11}, Q_{12}, and Q_{13}. The input impedance at the Y input (pins 6 and 11) is inversely proportional to the current i_1, or

$$R_i = \frac{K_2}{i_1} \tag{9.56}$$

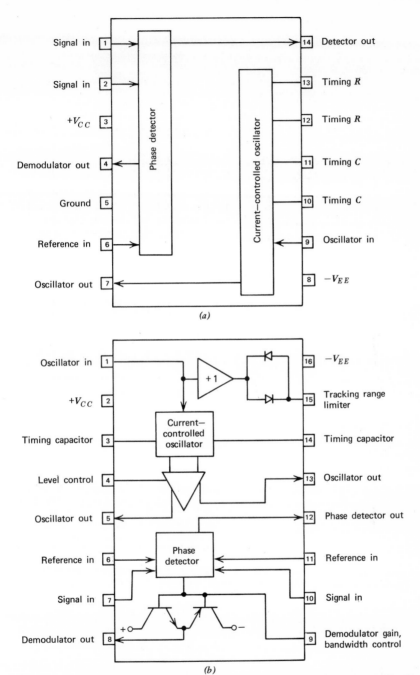

Figure 9.18 Functional diagrams of two PLLs. (*a*) Low-frequency PLL (HA-2820). (*b*) A high-frequency PLL (HA-2800).

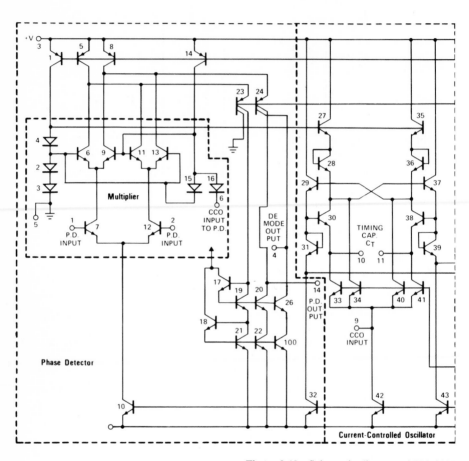

Figure 9.19 Schematic diagram of HA-2820

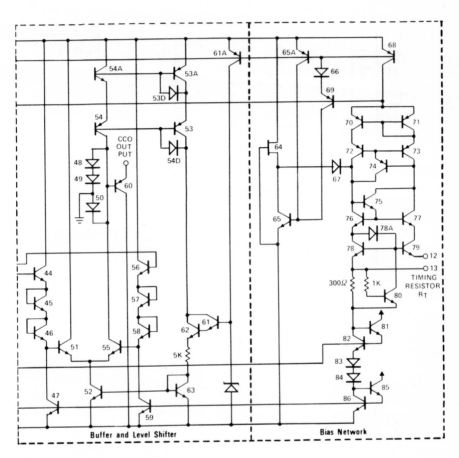

low-frequency phase-locked loop.

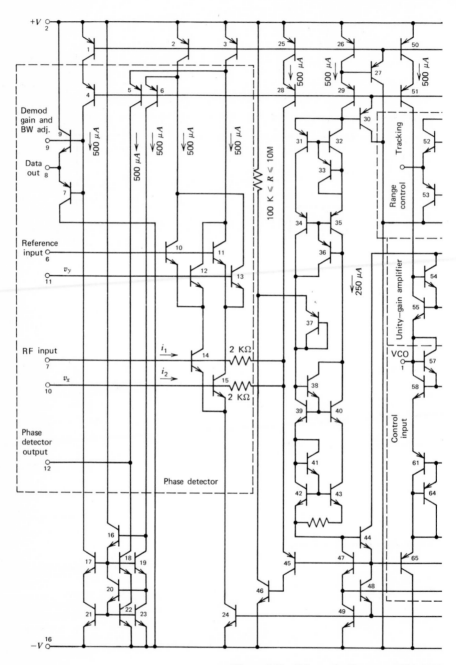

Figure 9.20 Schematic diagram of HA-2800

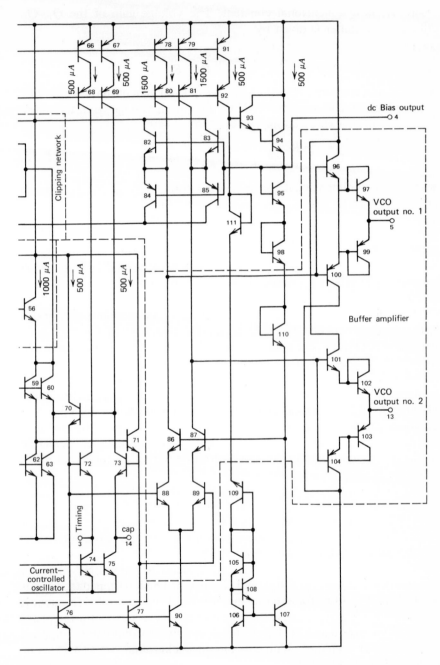

high-frequency phase-locked loop.

where K_2 is a dimensional constant. The voltage gain of the Q_{10}–Q_{13} difference amplifier is given by

$$A_v = \frac{v_{out}}{v_x} = A_I \frac{R_L}{R_i} = \frac{K_3}{R_i} \qquad (9.57)$$

where K_3 is a constant, A_I is the effective current gain of the stage, and R_L is the effective load impedance. Thus the output voltage from the phase detector can be expressed as

$$v_{out} = v_x \frac{K_3}{R_i} = v_x \frac{K_3}{K_2/i_1} = \frac{K_3}{K_2} K_1 v_x v_y = K_4 v_x v_y \qquad (9.58)$$

where K_4 is a dimensional constant. This output voltage is converted to an output current signal

$$i_{out} = K_5 v_x v_y \qquad (9.59)$$

by the Q_5 stage. Thus the current signal at pin 12 represents the analog multiplication between the two x and y inputs. A second current signal, proportional to the $v_x v_y$ product and buffered from the first, is available at pin 9.

The current-controlled oscillator is basically an astable multivibrator with coupling between the emitters of Q_{72} and Q_{73} via the C_T tuning capacitor. The quiescent collector currents of Q_{72} and Q_{73} are modulated by the dc current supplied at the low-impedance input of pin 1. An increase in the input current at this input is amplified by the Q_{59}–Q_{60} emitter-follower stages to produce an increase in the collector currents of Q_{72} and Q_{73}. When the timing capacitor C_T is in position, astable operation occurs because of the presence of positive feedback, where the loop gain around Q_{70} through Q_{73} is greater than unity. The frequency of oscillation is directly proportional to the input current externally supplied at pin 1. Thus we have

$$f_o = K i_{in} \qquad (9.60)$$

Two differential outputs are taken from the current-controlled oscillator at the bases of Q_{72} and Q_{73}. These signals are used to drive the differential input stage of the buffer amplifier (Q_{88} and Q_{89}). Complementary outputs are provided to pins 5 and 13 through the identical NPN–PNP stages of Q_{96} (Q_{101}), Q_{97} (Q_{102}), Q_{99} (Q_{103}), and Q_{100} (Q_{104}). The low-impedance levels at pins 5 and 13 allow for external monitoring of the controlled frequency, also making it possible to use these points as source signals for the phase detector via pins 6 and 11. Up to 10 mA of drive current can be supplied by this buffer amplifier. With this circuit background, we can examine several PLL applications.

FM Detector

A PLL connected to a receiver intermediate-frequency (IF) signal with its center frequency at the IF frequency will replace one or more IF stages and the discriminator stage, eliminating the cost, bulk, and lower reliability of IF and discriminator transformers. In broadcast receivers, PLLs can also be used to detect the stereo pilot signal and to demodulate subsidiary carrier authorization (SCA) broadcasts. In TV receivers, PLLs

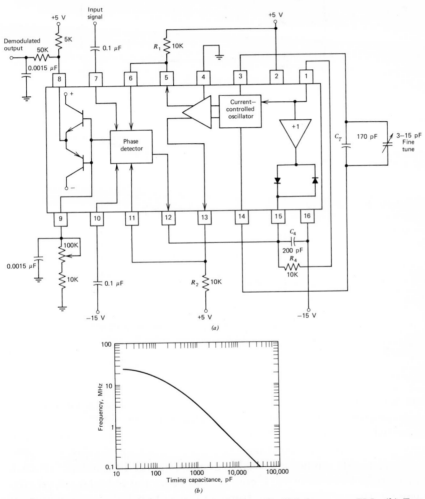

Figure 9.21 FM detector. (a) Circuit connections to the high-frequency PLL. (b) Free-running frequency versus timing capacitance. (c) Demodulated output swing versus input frequency deviation. (d) Tracking range versus input signal level.

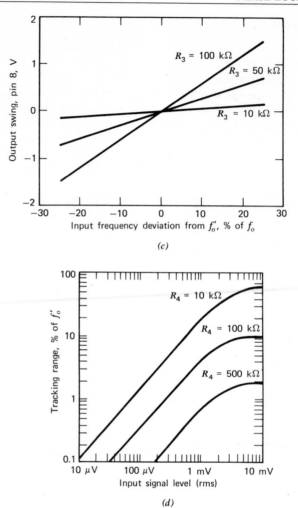

(c)

(d)

Figure 9.21 (cont.)

can serve in the IF, sound discriminator, synchronous separator, and color discriminator sections, in each case replacing LC tuned circuits. Operated as an IF receiver, the PLL is really acting as a notch filter to accomplish coherent detection.

The HA-2800 can be employed to demodulate the 10.7 MHz IF signal in commercial FM receivers. Figure 9.21a shows the external circuit connections to the HA-2800 for frequency-selective FM demodulation. If demodulation at a frequency other than 10.7 MHz is desired, the value of the required tuning capacitor C_T may be found from Figure 9.21b. The potentiometer R_3 provides a convenient method of adjusting the amplitude of the demodulated output signal. Variation of R_3 will typically change this amplitude from 3 to 30 mV rms when a 3 mV rms, 10.7 MHz input signal modulated at 1 kHz by a ±75 kHz deviation is applied. Figure 9.21c gives the variation of the demodulated output versus input frequency deviation for different R_3 settings of 10, 50, and 100 kΩ. A simple, passive, low-pass filter (R_4–C_4) is used between the phase detector and the current-controlled oscillator. Connecting one terminal of C_4 to the negative power supply as shown instead of to ground will provide better rejection of power-supply noise. Adjustment of R_4 gives independent control of the PLL's tracking range (see 9.21d). Setting R_4 to 10 kΩ gives a typical tracking range of 4 MHz for the same 10.7 MHz input signal described previously. The capture or acquisition range is independently controlled by the time constant of the low-pass filter. For the R_4–C_4 filter shown, the capture range may be approximated by

$$\Delta \omega_C \approx \pm \sqrt{\frac{\omega_T}{R_4 C_4}} \qquad (9.61)$$

where $\Delta \omega_T$ is the tracking range. The capture range is nominally 900 kHz for the test circuit of Figure 9.21a. The maximum nonlinearity of the FM detector is typically less than 1% at 100% modulation. Signal-to-noise ratio is greater than 40 dB for the 0.7% frequency deviation about the 10.7 MHz carrier. Total harmonic distortion is less than 1% for the 1 kHz modulation frequency; AM rejection exceeds 45 dB.

AM Detector

Two PLLs can be connected together as in Figure 9.22 to accomplish AM demodulation. The first PLL operates as a synchronous AM detector, and the loop locks onto the AM carrier frequency. The output frequency of the CCO is identical to the carrier without amplitude modulation. Demodulation is obtained by analog multiplication with the phase detector of the second PLL. After filtering, the difference-frequency component remains as the demodulated AM signal.

An additional phase-shift network is needed for AM detection because of the inherent 90° phase shift in the first PLL. By intentionally providing 90° of additional phase shift to the AM input, we ensure that both inputs to the phase detector are in phase. The phase detector then mixes the two inputs and produces an output directly proportional to the amplitude of the input signal. AM detection with PLLs can offer a higher degree of noise immunity than is typically achieved with conventional peak-detector-type AM demodulators. Such an improvement occurs because

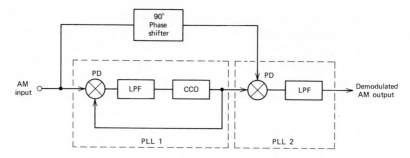

Figure 9.22 AM detector using phase-locked loops.

the PLL AM detection method is a coherent detection scheme in which the two compared signals are essentially averaged together.

A practical way of implementing an AM detector using a PLL is represented in Figure 9.23. A single PLL is used with a divide-by-2 counter inserted between the phase detector and the CCO. This counter forces the CCO frequency to be twice the input frequency. The frequencies of the two input signals to the exclusive-OR logic are $2f_i$ and f_i shifted

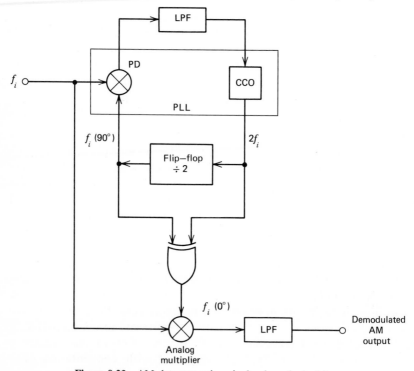

Figure 9.23 AM detector using single phase-locked loop.

by 90°. This logic function produces an output signal whose frequency equals the input frequency without any amplitude modulation. Beating this logic output signal against the original input, followed by low-pass filtering, produces the desired demodulated AM output. The major advantage of this AM detection scheme is that the flip-flop and exclusive-OR connection achieves a 90° phase shift for any input frequency without the need of special tuning methods.

Frequency Synthesis

Frequency synthesis, where many discrete frequencies can be synthesized from a given reference source, is another important application of the PLL. Since the loop is phase locked when PLL techniques are employed, the temperature-stability and drift characteristics of the synthesized frequencies will be identical to those of the reference source.

Frequency synthesis with PLLs utilizes a programmable counter (divide-by-N circuit) connected between the output of the current-controlled oscillator and the input to the phase detector. The counter divides the CCO's output frequency by a programmable integer, N. Figure 9.24a shows the circuit connections to the HA-2800 for a frequency synthesizer with $N = 4$ which is derived from a TTL flip-flop counter. When the loop is locked to an input signal of fundamental frequency f_s, the synthesized output frequency from the CCO is

$$Nf_s = 4f_s \qquad (9.62)$$

Values of N other than 4 can be easily employed by substituting standard logic connections to produce the desired count/divide sequence.

A 1170 pF tuning capacitor sets the free-running frequency of the current-controlled oscillator to 1.6 MHz. The divide-by-4 counter provides a 400 kHz as one input to the phase detector, and a 3 mV rms sinusoidal input signal acts as the other. Phase lock and tracking are possible for the fundamental frequency ($f_s = 400$ kHz) as well as a number of odd harmonics of f_s. Lock could be established for all odd harmonics up to and including the eleventh (4.4 MHz) with the test circuit of Figure 9.24a. Locking onto odd harmonics is to be expected, since the phase detector acts as an analog multiplier, forming the product between the sinusoidal input signal and the square-wave output from the CCO. The product of the odd-valued sine wave and the even-valued square wave is an odd-time function whose Fourier spectrum contains only odd terms. When locked to the kth harmonic of f_s, namely, f_{sk}, the output frequency at pin 5 is given by

$$f_o = N \frac{f_{sk}}{k} \qquad (9.63)$$

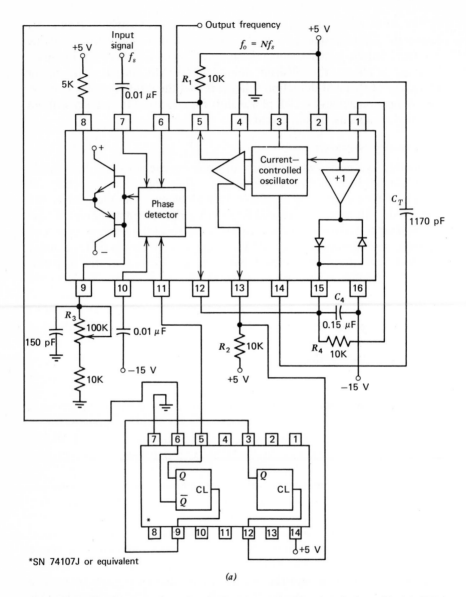

Figure 9.24 Circuit connections for frequency synthesizer, interfacing with (a) TTL counters, (b) ECL counters.

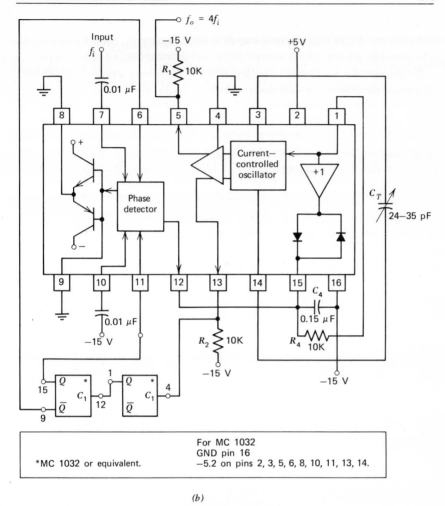

	For MC 1032
*MC 1032 or equivalent.	GND pin 16 −5.2 on pins 2, 3, 5, 6, 8, 10, 11, 13, 14.

(b)

Figure 9.24 (cont.)

The frequency limitations of certain TTL series should be considered when interfacing the high-frequency PLL. This consideration arises because at least one of the flip-flops in the divide-by-N chain must toggle at the CCO frequency, which can be set as high as 25 MHz. Frequencies in this range may not be attainable with the standard, 54/7400 TTL series. These considerations may necessitate using one of the higher-speed series, such as the MC2000/3000, the 54H/74H00, or the 54S/74S00.

For frequencies in the range of approximately 10 MHz and above, the amplitude of the CCO output may decrease to a level insufficient for driving TTL. In such cases, using a high-speed voltage comparator between the CCO and the first flip-flop will "square up" the CCO output and provide an improved wave shape for driving the TTL flip-flops.

Employing ECL presents another possible but more expensive solution to this high-frequency constraint. Figure 9.24b shows the same $N = 4$ PLL frequency synthesizer using ECL flip-flops. Lock was easily achieved with a 50 mV rms sinusoidal input. The C_T tuning capacitor adjusted the CCO frequency between 20 and 15 MHz. The output amplitude of the signal from the CCO was sufficiently large to clock the first ECL flip-flop to approximately 22 MHz. For larger frequencies it may be necessary to utilize a high-frequency shaping amplifier at this point to "square up" the output signal from the CCO.

A second counter may be added to the previous circuits to synthesize numerous output frequencies which are fractionally related to the reference input frequency. Figure 9.25a is a block diagram for such a scheme. The reference-frequency input f_r is divided by an integer M and applied as the input to the PLL. The PLL then functions as previously explained, but it now has an input of f_r/M. The synthesized frequency at the output is

$$f_o = \frac{Nf_r}{M} \tag{9.64}$$

Therefore, by programming M and N, any of a large number of frequencies fractionally related to f_r can be synthesized, each having the stability of the single reference frequency. The circuit of Figure 9.25b reveals connections to the HA-2800 for fractional frequency synthesis. This circuit is identical to those of Figure 9.24 except for the divide-by-M counter stages at the input. The signal level from the last flip-flop in this

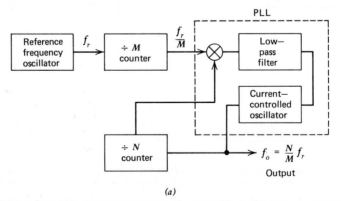

(a)

Figure 9.25 Synthesizing fractional frequencies. (*a*) Block-diagram representation. (*b*) Circuit implementation with high-frequency PLL.

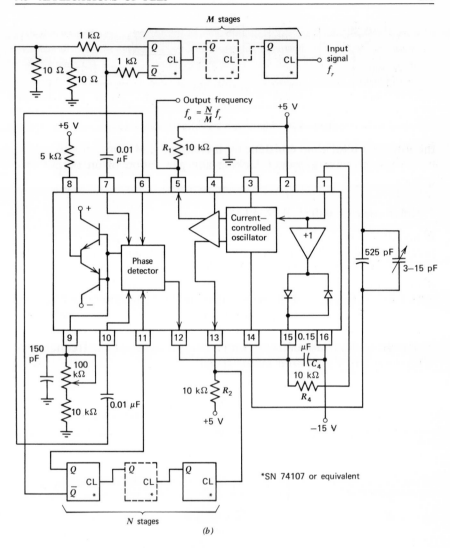

Figure 9.25 (cont.)

chain should be attenuated to a level less than 400 mV rms before being applied to the phase detector (pins 7 and 10). Large voltages here will exceed the input-voltage range of the phase detector, thereby causing saturation and preventing the loop from locking on to f_r. This circuit produces an output signal at pin 5 whose frequency is (N/M) times f_r when the loop is locked. Locking of the loop on to harmonics of f_r may occur for the same reasons as discussed for Figure 9.24. Adjustment of

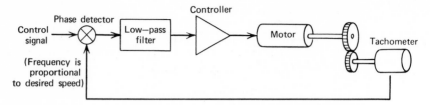

Figure 9.26 Motor speed controller using PLL.

the low-pass filter time constant by increasing R_4–C_4 presents the most direct way to ensure proper locking and generation of an output frequency of Nf_r/M.

Motor Speed Control With PLLs

Many electromechanical systems, such as magnetic tape drives, require precise speed control, particularly during start and stop operations. Figure 9.26 illustrates the incorporation of a motor control within a phase-locked loop. A controller block has been added to boost the power-drive capability of the phase-detector–low-pass-filter combination. A tachometer is attached to the motor shaft, producing an ac signal whose frequency is directly proportional to the speed of the motor. This signal is compared in the phase detector with a control signal whose frequency is proportional to the desired motor speed.

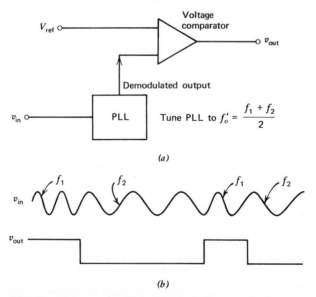

Figure 9.27 FSK demodulator using PLL. (a) Block diagram. (b) System wave shapes.

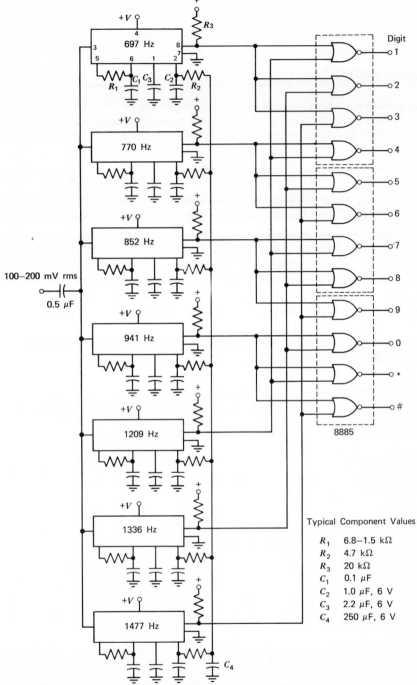

Figure 9.28 Touch-tone encoder using PLLs. This application requires AM-detector-type PLLs as described in Figure 9.23. (Courtesy of Signetics Corporation, a division of Corning Glass Works.)

375

Data Modems

In transmitting digital data over telephone lines, the ones and zeros are often encoded as two different audio tone frequencies. This is known as frequency shift keying (FSK). A PLL tuned to the middle of the two input frequencies may be used to decode these tones, as diagrammed in Figure 9.27. Since the demodulated output from the PLL is proportional to the input frequency, a voltage comparator will discriminate these two frequencies. Other modem systems utilize phase modulation of a single audio frequency, which can also be detected by a PLL in terms of spikes at the filter output. A more sophisticated example of a modem is a Touch-Tone® encoder, which will ultimately become a part of every telephone. The number "dialed" is converted to a combination of two particular frequencies. The function of the encoder modem is to reconstruct the original number by detecting which two frequencies are present. The most obvious way of accomplishing this detection is with tuned circuits, each tuned to a different frequency to act as filter. However, monolithic PLLs are ideally suited for this filter applications. The scheme in Figure 9.28 has seven PLLs acting as tone encoders for a common input signal. The PLL outputs drive 12 NOR gates to produce the 10 digital outputs plus the * and # indexes.

Tracking Filter

As a final application, consider using a PLL, an active filter, and an analog multiplexer in a system to form a tracking filter like that in Figure 9.29. A variable resistor is used to sweep the CCO frequency until the PLL acquires the input signal. Once acquired, the PLL tracks and filters the input signal by charging the filter capacitance through the analog multiplexer. With this tracking-filter arrangement it is possible to achieve very high effective Q's with a high signal-to-noise ratio.

9.9 PLL PARAMETERS

The question most often asked about certain phase-locked loop parameters appearing on a specification sheet is, What do they mean, and how can I interpret them for my application? Let us examine several of the most misunderstood parameters.

Phase Detector

The *Input Voltage* entry under the absolute maximum ratings heading represents an upper limit for amplitude of the input signal to the phase

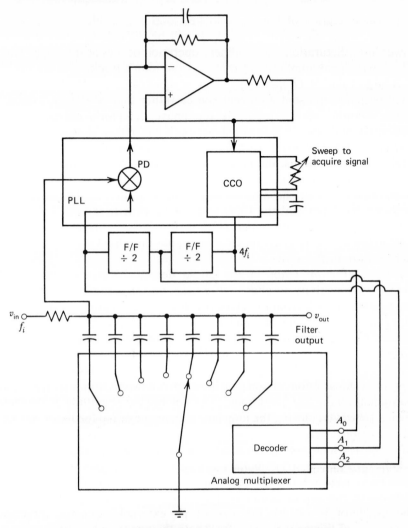

Figure 9.29 Tracking filter using PLL and analog multiplexer.

detector. This rating is significant in TTL frequency-synthesizer applications in which the input signal (or signals) is obtained from the output of a counter.

The typical *Input Impedance* is between the two inputs to the phase detector [pins 7 and 10 of Figure 9.18*b*]. It is not the impedance from pin 7 or pin 10 to ground. This distinction is important when the PLL is connected to differential signal sources. The value of *input impedance* is considered when choosing the size of the input coupling capacitors.

Linear operation of the phase-locked loop is possible over the *Input Voltage Range* specified. Phase lock will be difficult to achieve at the lower limit. Saturation and nonlinear operation occurs at the upper limit. The input-signal amplitude also affects the loop's tracking range.

A large value of *Output Impedance* is significant only to illustrate that such outputs represent a current source rather than a voltage source.

The *Output-Offset Current* is the phase detector's output current measured with the loop opened between the phase detector and the low-pass filter. The test is made under zero input-signal conditions. Under closed-loop conditions, the offset current will be reduced by a factor of K_d, which is the phase detector's conversion gain (see next entry). This, in turn, causes a small phase offset within the loop; that is, the phase angle between the input and the CCO will not be exactly 90° ($\pi/2$ rads) but $\pi/2 \pm I_{offset}/K_d$ rads. Any temperature drift in offset current will cause a drift in the closed-loop center frequency; but this effect is included in the temperature-drift curve.

The *Conversion Gain* of the phase detector K_d is a measure of detector's sensitivity to phase differences between the input signal and the signal fed back from the current-controlled oscillator. From the data entry here, we see that the phase detector will typically give an output signal of 50 μA for every radian of phase difference. Since the characteristic of the phase detector is similar to a cosine curve, the conversion gain is measured as the slope of the characteristic as it passes through 90°. Specific values of conversion gain in an open-loop environment are highly dependent on the test conditions, most notably the amplitude of the input signal. However, closing the loop and utilizing negative feedback reduces K_d variations.

Current-Controlled Oscillator

The free-running or center frequency f'_o (unlocked-loop condition) of the oscillator is set by the size of an external capacitor. Practical limitations of this capacitance affect both the upper and lower frequency ranges. Highly stable, low-leadage capacitors are rare above 10 μF, which limits the lower end of the frequency range to approximately 10 Hz. The *Maximum Frequency* is limited by internal and external stray capacitances to typically 30 MHz.

The *Frequency Drift* is the change in f'_o under open-loop conditions per degree C change in temperature.

The *Frequency Change with Supply Voltage* is the percentage f'_o change (open-loop) accompanying a one-volt change in the power-supply voltages.

A low value of *Input Resistance* implies that the CCO is sensitive to a current (as opposed to a voltage) input signal. This resistance should be considered in the design of the low-pass filter.

The *Input Open-Circuit Voltage* represents a dc reference level at the CCO input. This level should be taken into consideration if an active, low-pass filter is used here or if an external ramp current is applied to sweep the CCO frequency.

The *Clipping Level* results from two parallel diodes connected to the output of the unity-gain amplifier (see Figure 9.8). The clipping level is referenced about the *input open-circuit voltage*. The clipping network provides independent control of the tracking range through selection of a single resistor.

The *Conversion Gain K_o* specifies the sensitivity of the oscillator's frequency to change in accordance with changes in the input current from the phase detector and low-pass filter. We have seen that $1 \mu A$ will typically change the CCO frequency by 1%. This *conversion gain* is simply the slope of oscillator frequency versus input current curve.

The *Output Voltage and Output Rise and Fall Times* are particularly significant when interfacing the CCO output with a digital counter.

Closed-Loop Characteristics

The *Conversion Gain* is the ratio of output change to input change in a portion of the loop. In the phase detector, this is the ratio of output signal change to a change in phase angle between the two inputs (volts per radian or microamperes per radian). Since the phase-detector characteristic is generally a cosine curve, the gain is measured for small excursions about the 90° point. In the oscillator this is the ratio of output frequency change to an input change ($\% \Delta f/V$ or $\% \Delta f/\mu A$).

The *Loop Gain* is the product of the conversion gains of the phase detector and the current controlled oscillator (i.e., $K_d K_o$ in units of $\% \Delta f/rad$). The loop gain reflects the ratio of change in input (and oscillator) frequency to the change in phase shift between input and oscillator. Therefore, loops with higher gains will hold the phase relationship closer to 90° for a given input frequency change, and conversely, will have a broader tracking range.

The *Capture or Acquisition Range* is the range of input frequency about f'_o under which a PLL that is initially unlocked, will become locked. This range is narrower than the normal tracking range, and it is a function of both the loop-filter characteristics and the input amplitude.

The *Tracking or Lock Range* is the range of input frequency about f'_o under which a PLL, once locked, will remain locked. This is a function of loop gain, since the phase-detector output is bounded and can drive the oscillator only over a certain frequency range. The tracking range may be controlled, if desired, by limiting the input signal to the oscillator.

9.10 CONCLUSIONS

In this chapter we described the basic theory of phase-locked loop operation and developed a linear model for predicting the performance of a typical system. The effects of several typical low-pass filters on system stability and response were described using root-locus and steady-state error-analysis techniques. The circuit operation of a PLL was discussed, as a basis for a better interpretation of the performance characteristics of a PLL device. Methods were presented for obtaining useful parameters of a linear PLL model from a device's performance curves. Considerations involved in making external connections to a PLL were illustrated with typical applications. Specific applications considered were FM and AM detection, frequency synthesis, data modems, and motor speed control.

BIBLIOGRAPHY

1. Cheng, David K., *Analysis of Linear Systems*, Addison-Wesley, Reading, Mass., 1961.
2. Chestnut, Harold, and Robert W. Mayer, *Servomechanisms and Regulating System Design*, Wiley, New York, 1959.
3. D'Azzo, J. J., and C. H. Houpis, *Feedback Control System Analysis and Synthesis*, McGraw-Hill, New York, 1966.
4. De Bellescize, H., "La Reception Synchrone," *Onde* (*Electronics*) **11**, (June 1932), pp. 230–240.
5. Gardner, F. M., *Phaselock Techniques*, Wiley, New York, 1966.
6. Gardner, Murry F., and John L. Barnes, *Transients in Linear Systems*, Wiley, New York, 1963.
7. Grebene, Alan B., *Analog Integrated Circuit Design*, Van Nostrand-Reihhold, New York, 1972.
8. Gruen, W. J., "Theory of AFC Synchronization," *Proc. IRE*, **41** (August 1953), pp. 1043–1048.
9. Kuo, B. C., *Automatic Control Systems*, Prentice-Hall, Englewood Cliffs, N.J., 1962.
10. Kuo, B. C., *Analysis and Synthesis of Sampled-Data Control Systems*, Prentice-Hall, Englewood Cliffs, N.J., 1963.
11. Lathi, B. P., *Random Signals and Communication Theory*, International Textbook, Scranton, Pa., 1968.
12. Motorola Semiconductor Products Inc., "Linear Integrated Circuits Data Book," Motorola, Inc., December 1972.

13. Motorola Semiconductor Products Inc., "TTL Integrated Circuits Data Book," Motorola, Inc., May 1972.

14. Rey, T. J., "Automatic Phase Control: Theory and Design," *Proc. IRE*, **48** (October 1960), pp. 1760–1771. Corrections in *Proc. IRE*, March 1961, p. 590.

15. Richman, D., "APC Color Sync. for NTSC Color Television," *IRE Convention Record*, 1953, part 4.

16. Viterbi, A. J., "Acquisition and Tracking Behavior of Phase Locked Loops," JPL External Publication No. 673, July 14, 1959.

17. Viterbi, A. J., *Principles of Coherent Communication*, McGraw-Hill, New York, 1966.

APPENDIX A

DERIVATION OF OUTPUT CONDUCTANCE

The output conductance at the collector of a transistor is of considerable interest in the design of constant-current circuits. We will derive a useful expression for this conductance associated with a transistor that has resistors in the base and emitter legs, as in Figure A.1. Using the common-base, hybrid model we can write

$$i_o = h_{fb}i_e + h_{ob}v_{cb} \tag{A.1}$$

$$v_{cb} = v_o - R_B(i_e + i_o) \tag{A.2}$$

$$R_B(i_e + i_o) + h_{rb}v_{cb} + i_e(h_{ib} + R_E) = 0 \tag{A.3}$$

After simplification of these equations, we have

$$i_o(1 + R_B h_{ob}) = i_e(h_{fb} - h_{ob}R_B) + h_{ob}v_o \tag{A.4}$$

$$i_e = -\frac{R_B(1 - h_{rb})i_o}{R_E + h_{ib} + R_B(1 - h_{rb})} - \frac{h_{rb}v_o}{R_E + h_{ib} + R_B(1 - h_{rb})} \tag{A.5}$$

Two approximations are possible when the typical parameters from Figure 2.11 are utilized in Equations A.1 through A.5. We see that $h_{rb} \ll 1$ and $h_{ob}R_B \ll 1$ for practical resistors fabricated in monolithic ICs. Based on these two approximations, Equations A.4 and A.5 reduce to

$$i_o \approx h_{fb}i_e + h_{ob}v_o \tag{A.6}$$

$$i_e \approx \frac{-R_B i_o}{R_E + h_{ib} + R_B} - \frac{h_{rb}v_o}{R_E + h_{ib} + R_B} \tag{A.7}$$

Combining and simplifying the two previous equations gives

$$i_o[R_E + h_{ib} + R_B(1 + h_{fb})] \approx v_o[h_{ob}(R_E + R_B + h_{ib}) - h_{rb}h_{fb}] \tag{A.8}$$

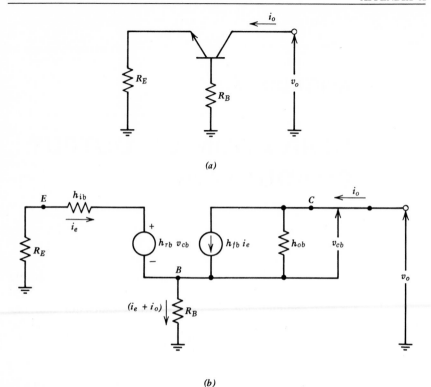

(a)

(b)

Figure A.1 Finding output conductance of transistor circuit, looking into collector. (a) Complete circuit. (b) Common-base hybrid equivalent circuit model.

Rearranging, we have

$$i_o = v_o \left[\frac{h_{ob}(R_E + R_B + h_{ib}) - h_{rb}h_{fb}}{R_E + h_{ib} + R_B(1 + h_{fb})} \right] \tag{A.9}$$

$$i_o = v_o \left[\frac{h_{ob}[R_E + h_{ib} + R_B(1 + h_{fb}) - h_{fb}R_B]}{R_E + h_{ib} + R_B(1 + h_{fb})} - \frac{h_{rb}h_{fb}}{R_E + h_{ib} + R_B(1 + h_{fb})} \right] \tag{A.10}$$

$$\therefore \ g_{out} = \frac{i_o}{v_o} = h_{ob}\left[1 - \frac{h_{fb}R_B}{h_{ib} + R_E + R_B(1 + h_{fb})} \right] - \frac{h_{fb}h_{rb}}{R_E + h_{ib} + R_B(1 + h_{fb})} \tag{A.11}$$

For most integrated circuit applications,

$$R_B(1 + h_{fb}) \ll h_{ib} + R_E \tag{A.12}$$

and g_{out} reduces to

$$g_{out} = h_{ob}\left(1 - \frac{h_{fb}R_B}{h_{ib} + R_E} \right) - \frac{h_{fb}h_{rb}}{h_{ib} + R_E} \tag{A.13}$$

APPENDIX B

DERIVATION OF h_{ob} CANCELLATION TECHNIQUE

Refer to the h_{ob} neutralization circuit of Figure 3.31b. It was shown in Equation 3.52 that the output conductance for the stacked current mirror is dominated by the h_{ob} term. It was also shown that h_{rb} terms were negligible compared with h_{ob} effects.

Although our derivation here is not intended to be rigorously correct, it indicates the underlying principles in the h_{ob} cancellation technique. For simplicity, let us assume

$$h_{fb} \approx -1 \quad \text{and} \quad h_{rb} \approx 0 \qquad (B.1)$$

Applying these assumptions allows the modeling of Figure 3.31b, as in Figure B.1. Considering all currents and voltages to be ac quantities, we can write

$$i_{in} \approx i_a + i_{e3} \qquad (B.2)$$

and

$$v_{in} \approx i_a\left(\frac{1}{h_{ob3}} + h_{ib1} + h_{ib2} + R\right) + (i_a h_{ib1} - i_{e3} h_{ib3})h_{ob4}(h_{ib2} + R) \qquad (B.3)$$

From the current relationships given in Equation 3.54, we can conclude

$$h_{ib1} = h_{ib2} = h_{ib3} = h_{ib4} = 2h_{ib5} = h_{ib} \qquad (B.4)$$

Solving for i_{e4}, we have

$$i_{e4} \approx i_a(1 + h_{ib}h_{ob4}) - i_{e3}h_{ib}h_{ob4} \qquad (B.5)$$

385

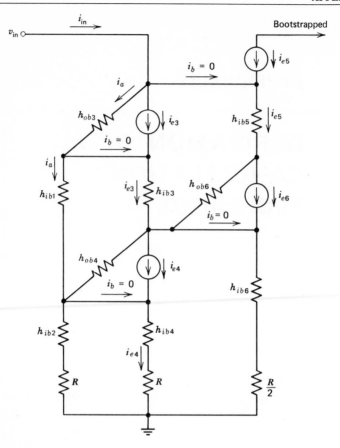

Figure B.1 Simplified equivalent circuit model of Figure 3.31b for showing h_{ob} neutralization technique.

Also, we can write

$$i_{e4} \approx i_{e3}\{1 + h_{ib}h_{ob4} + h_{ib}h_{ob6}[h_{ob4}(h_{ib} + R) + 1]\}$$
$$- i_a\{h_{ib}h_{ob4} + [(2h_{ib} + R) + h_{ib}h_{ob4}(h_{ib} + R)]h_{ob6}\}$$
$$- \left(i_{e5}\frac{h_{ib}}{2} - v_{in}\right)h_{ob6} \tag{B.6}$$

To simplify the foregoing equations, let

$$\frac{1}{h_{ob3}} \gg h_{ib} \tag{B.7}$$

$$h_{ib}h_{ob4} \ll 1 \tag{B.8}$$

$$h_{ob6}(h_{ib} + R) \ll 1 \tag{B.9}$$

$$i_{e5}\frac{h_{ib}}{2} \ll v_{in} \tag{B.10}$$

Applying these assumptions allows Equations B.3, B.5, and B.6 to be simplified as follows:

$$v_{in} \approx i_a\left(\frac{1}{h_{ob\,3}} + h_{ib} + R\right) - i_{e\,3}h_{ib}h_{ob\,4}(h_{ib} + R) \tag{B.11}$$

$$i_{e4} \approx i_a - i_{e3}h_{ib}h_{ob4} \tag{B.12}$$

$$i_{e4} \approx i_{e3}(1 + h_{ib}h_{ob6})$$
$$\quad - i_a\{h_{ib}h_{ob4} + h_{ob6}R + h_{ib}h_{ob6}[2 + h_{ob4}(h_{ib} + R)]\} + v_{in}h_{ob6} \tag{B.13}$$

For further simplification, assume

$$h_{ib} + R \ll \frac{1}{h_{ob3}} \tag{B.14}$$

$$h_{ib}h_{ob6} \ll 1 \tag{B.15}$$

$$h_{ob4}(h_{ib} + R) \ll 2 \tag{B.16}$$

Equations B.11 and B.13 further simplify to

$$v_{in} \approx \frac{i_a}{h_{ob3}} - i_{e3}h_{ib}h_{ob4}(h_{ib} + R) \tag{B.17}$$

$$i_{e4} \approx i_{e3} - i_a[h_{ib}h_{ob4} + h_{ob6}(2h_{ib} + R)] + v_{in}h_{ob6} \tag{B.18}$$

Substituting Equation B.18 into B.12 gives

$$h_{ob6}v_{in} = i_a[1 + h_{ib}h_{ob4} + h_{ob6}(2h_{ib} + R)] - i_{e3}(1 + h_{ib}h_{ob4}) \tag{B.19}$$

Considering, together with the previous assumptions, that

$$h_{ob6}(2h_{ib} + R) \ll 1 \tag{B.20}$$

we have

$$h_{ob6}v_{in} \approx i_a - i_{e3} \tag{B.21}$$

Substitution of the previous equation into Equation B.17 gives

$$v_{in}[1 - h_{ob6}h_{ob4}h_{ib}(h_{ib} + R)] \approx \frac{i_a}{h_{ob3}}[1 - h_{ob3}h_{ob4}h_{ib}(h_{ib} + R)] \tag{B.22}$$

Since

$$h_{ob6}h_{ob4}h_{ib}(h_{ib} + R) \ll 1 \tag{B.23}$$

and

$$h_{ob3}h_{ob4}h_{ib}(h_{ib} + R) \ll 1 \tag{B.24}$$

Equation B.22 reduces to

$$v_{\text{in}} \approx \frac{i_a}{h_{ob3}}$$

(B.25)

Combining this result with Equation B.21 gives

$$i_3 \approx v_{\text{in}}(h_{ob3} - h_{ob6})$$

(B.26)

Now from Equation B.2, we have the desired result

$$i_{\text{in}} \approx h_{ob3}v_{\text{in}} + (h_{ob3} - h_{ob6})v_{\text{in}}$$

(B.27)

$$g_{\text{in}} = \frac{i_{\text{in}}}{v_{\text{in}}} \approx 2h_{ob3} - h_{ob6}$$

(B.28)

In addition to exhibiting near infinite input resistance, the circuit of Figure 3.31b does not introduce first–order base current errors to its differential current drive. If we assume $\beta \gg 1$ for all transistors, we can write

$$I_{C1} \approx I_0 - \frac{I}{\beta_3} \approx I + \frac{I}{\beta_4}$$

(B.29)

or alternatively,

$$I \approx I_0 - \frac{I}{\beta_3} - \frac{I}{\beta_4}$$

(B.30)

The emitter and collector currents of Q_3 are

$$I_{E3} = I_{C4} + I_{B6} \approx I - \frac{I}{\beta_4} + \frac{2I}{\beta_6}$$

(B.31)

$$I_{C3} = I_{E3} - I_{B3} \approx I_{E3} - \frac{I}{\beta_3}$$

(B.32)

After some algebraic reduction, we can write an expression for I_{in} as

$$I_{\text{in}} \approx I_{C3} + \frac{2I}{\beta_5} \approx I_0 - 2I\left(\frac{1}{\beta_3} + \frac{1}{\beta_4} - \frac{1}{\beta_5} - \frac{1}{\beta_6}\right)$$

(B.33)

If good β matching exists among Q_3, Q_4, Q_5, and Q_6, we have

$$I_{\text{in}} \approx I_0$$

(B.34)

and the base-current errors are neutralized.

COMPLETE
BIBLIOGRAPHY

1. Ahmed, H., and P. J. Spreadbury, *Electronics for Engineers, An Introduction,* Cambridge University Press, London, 1973.

2. Angelo, E. James, Jr., *Electronics: BJTs, FETs, and Microcircuits,* McGraw-Hill, New York, 1969.

3. Barna, Arpad, *Operational Amplifiers,* Wiley, New York, 1971.

4. Belove, Charles, Harry Schachter, and Donald L. Schilling, *Digital and Analog Systems, Circuits, and Devices: An Introduction,* McGraw-Hill, New York, 1973.

5. Cheng, David K., *Analysis of Linear Systems,* Addison-Wesley, Reading, Mass., 1961.

6. Chestnut, Harold, and Robert W. Mayer, *Servomechanisms and Regulating System Design,* Wiley, New York, 1959.

7. Chirlian, Paul M., *Integrated and Active Network Analysis and Synthesis,* Prentice-Hall, Englewood Cliffs, N.J., 1967.

8. Connelly, J. A., N. C. Currie, and D. S. Bonnet, "Op Amp Has Sixteen-Step Digital Gain Control," *Electronic Design News,* **19,** No. 9 (May 5, 1974), pp. 75–77.

9. D'Azzo, J. J., and C. H. Houpis, *Feedback Control System Analysis and Synthesis,* McGraw-Hill, New York, 1966.

10. De Bellescize, H., "La Reception Synchrone," *Onde (Electronics),* **11** (June 1932), pp. 230–240.

11. Deboo, Gordon J., and Clifford N. Burros, *Integrated Circuits and Semiconductor Devices: Theory and Application,* McGraw-Hill, New York, 1971.

12. Dooley, D. J., "Applying a Monolithic 10-Bit D/A Converter," Application note for mono DAC-02, Precision Monolithics, Santa Clara, Calif.

13. Eimbinder, Jerry, *Semiconductor Memories,* Wiley, New York, 1971.

14. Fitchen, Franklin C., *Electronic Integrated Circuits and Systems,* Van Nostrand-Reinhold, New York, 1970.

15. Gardner, F. M., *Phaselock Techniques,* Wiley, New York, 1966.

16. Gardner, Murry F., and John L. Barnes, *Transients in Linear Systems,* Wiley, New York, 1963.

17. Graeme, Jerald G., Gene E. Tobey, and Lawrence P. Huelsman, *Operational Amplifiers: Design and Applications*, McGraw-Hill, New York, 1971.

18. Graeme, Jerald G. *Applications of Operational Amplifiers: Third Generation Techniques*, McGraw-Hill, New York, 1973.

19. Gray, Paul E., and Campell L. Searle, *Electronic Principles: Physics, Models, and Circuits*, Wiley, New York, 1969.

20. Grebene, Alan B., *Analog Integrated Circuit Design*, Van Nostrand-Reinhold, New York, 1972.

21. Gruen, W. J., "Theory of AFC Synchronization," *Proc. IRE*, **41** (August 1953), pp. 1043–1048.

22. Hilburn, John L., and David E. Johnson, *Manual of Active Filter Design*, McGraw-Hill, New York, 1973.

23. Hoeschele, David, Jr., *Analog-to-Digital, Digital-to-Analog Conversion Techniques*, Wiley, New York, 1968.

24. Huelsman, L. P., *Active Filters: Lumped, Distributed, Integrated, Digital and Parametrics*, McGraw-Hill, New York, 1970.

25. Hunter, Lloyd P., *Handbook of Semiconductor Electronics*, McGraw-Hill, New York, 1970.

26. Jones, Don, and Robert W. Webb, "Chopper-Stabilized Op Amp Combines MOS and Bipolar Elements on One Chip," *Electronics*, **46,** No. 20 (September 27, 1973), pp. 110–114.

27. Kuo, B. C., *Automatic Control Systems*, Prentice-Hall, Englewood Cliffs, N.J., 1962.

28. Kuo, B. C., *Analysis and Synthesis of Sampled-Data Control Systems*, Prentice-Hall, Englewood Cliffs, N.J., 1963.

29. Lathi, B. P., *Random Signals and Communication Theory*, International Textbook, Scranton, Pa., 1968.

30. Le Page, W. R., and S. Seely, *General Network Analysis*, McGraw-Hill, New York, 1952.

31. Meyer, Charles S., David K. Lynn, and Douglas J. Hamilton, *Analysis and Design of Integrated Circuits*, McGraw-Hill, New York, 1968.

32. Melen, Roger, and Harry Garland, *Understanding IC Operational Amplifiers*, Sams, Indianapolis, 1971.

33. Millman, Jacob, and Christos C. Halkias, *Electronics Devices and Circuits*, McGraw-Hill, New York, 1967.

34. Millman, J., and C. C. Halkias, *Integrated Electronics: Analog and Digital Circuits and Systems*, McGraw-Hill, New York, 1972.

35. Millman, J., and Herbert Taub, *Pulse, Digital, and Switching Waveforms*, McGraw-Hill, New York, 1965.

36. Morris, Robert L., and John R. Miller, *Designing With TTL Integrated Circuits*, McGraw-Hill, New York, 1971.

37. Motchenbaucher, C. D., and F. C. Fitchen, *Low Noise Electronic Design*, Wiley, New York, 1973.

38. Motorola Semiconductor Products Inc., "Linear Integrated Circuits Data Book," Motorola, Inc., December 1972.

39. Motorola Semiconductor Products Inc., "TTL Integrated Circuits Data Book," Motorola, Inc., May 1972.

40. Peatman, John B., *The Design of Digital Systems*, McGraw-Hill, New York, 1972.

41. Pierce, J. F., and T. J. Paulus, *Applied Electronics*, Merrill, Columbus, Ohio, 1972.

42. Pierce, J. F., *Semiconductor Junction Devices*, Merrill, Columbus, Ohio, 1967.

43. Pierce, J. F., *Transistor Circuit Theory and Design*, Merrill, Columbus, Ohio, 1963.

44. RCA Inc., "Linear Integrated Circuits," RCA Corp., Technical Series IC-42, 1970.

45. RCA Inc., "Linear Integrated Circuits and MOS Devices," RCA Corp. No. SSD-202, 1972.

46. Rey, T. J., "Automatic Phase Control: Theory and Design," *Proc. IRE*, **48** (October 1960), pp. 1760–1771. Corrections in *Proc. IRE*, March 1961, p. 590.

47. Richman, D., "APC Color Sync. for NTSC Color Television," *IRE Convention Record*, 1953, part 4.

48. Ryder, John D., *Electronic Fundamentals and Applications*," Prentice-Hall, Englewood Cliffs, N.J., 1970.

49. Schilling, Donald L., and Charles Belove, *Electronic Circuits: Discrete and Integrated*, McGraw-Hill, New York, 1968.

50. Schwartz, Seymour, *Integrated Circuit Technology*, McGraw-Hill, New York, 1967.

51. Sheingold, Daniel H., *Analog-Digital Conversion Handbook*, Analog Devices Inc., Norwood, Mass., 1972.

52. Solomon, J. E., W. R. Davis, and P. L. Lee, "A Self-Compensated Monolithic Operational Amplifier with Low Input Current and High Slew Rate," IEEE International Solid State Circuits Conference, Philadelphia, February 19, 1969, pp. 14–15.

53. Stewart, Harry E., *Engineering Electronics*, Allyn & Bacon, Boston 1969.

54. Strauss, Leonard, *Wave Generation and Shaping*, McGraw-Hill, New York, 1970.

55. Su, Kendell L., *Active Network Synthesis*, McGraw-Hill, New York, 1965.

56. Viterbi, A. J., "Acquisition and Tracking Behavior of Phase Locked Loops," JPL External Publication No. 673, July 14, 1959.

57. Viterbi, A. J., *Principles of Coherent Communication*, McGraw-Hill, New York, 1966.

58. Warner, Raymond M., Jr., and James N. Fordemwalt, *Integrated Circuits: Design Principles and Fabrication*, McGraw-Hill, New York, 1965.

INDEX

Acceptor impurity, 8
Access time, *see* Multiplexer
  parameters
Accuracy, 292, 315, 317
Acquisition range, 367, 379
Active filters, 158, 174, 176,
  179, 260, 350, 353, 376
Address decoder, 322
Admittance cancellation, 94
AEDCAP, 50
Air isolation, 11
Alpha, 42
Aluminum, 5, 11, 15, 19, 29,
  38, 40, 41
AM detector, 366, 368, 369,
  375
AM rejection, 367
Analog-to-digital converter,
  113, 119, 138, 158, 162, 263,
  264, 287, 317
Analog-voltage-to-digital-
  pulse converter, 136, 137
Analog computer, 158
Analog multiplier, 103, 106,
  162, 358, 364, 367
Analog signal range, 275
Analog switch, 108, 109, 113,
  229, 233, 253, 254
 sixteen channel, 271
Antilog amplifier, 160, 161
Antimony, 8
Argon, 11
Arsenic, 8, 10
Astable multivibrator, 107,
  137, 165, 166, 168, 169, 332,
  364

Automatic gain control, 235

Balanced-bridge configuration,
  69, 70
Bandwidth, 180, 183, 191, 200,
  219, 223, 224, 226, 248
 control, *see* Frequency compensation
Base, formation in transistors,
  18, 26
 resistivity, 21
 transport factor, 31
 width, 21, 25
Base-collector junction, 16,
  31, 47
Base-current cancellation, 94,
  96
Base-emitter junction, 47
Beat frequency, 338
Beta, 19, 30, 31, 42, 44, 48,
  49, 94, 123
Bias current, 60, 87, 91, 94,
  123, 126, 128, 142, 143, 157,
  160, 169, 193-197, 202-204,
  219, 220 223-226, 235, 239,
  254, 266, 267, 281
 network, 67, 122
Binary-coded decimal, 308
Binary-weighted, currents, 304
 resistor network, 113, 114,
  297, 304
Bootstrapping, 91, 97, 100, 101
Boron, 8
Break-before-make-delay, *see*
  Multiplexer parameters
Buffer amplifier, 124, 252,

Voltage gain, 126, 128, 181
Voltage regulator, 162-164

Wafer, 1
Window detector, 134

X-Y recorder, 267, 268
Xylene, 5

Zero-crossing detector, 131-133